Food Technology Lab Manual

The Author

Dr. Gboyega Oralugbagbe worked as Professor of Food Science and Nutrition at University of Nigeria, Nsukka. Presently Dr. Oralugbagbe is working as Food Consultant in the Area of Food Processing, Food Safety, Quality Systems, HACCP, Sanitation and GMP compliance.

Food Technology Lab Manual

Dr. Gboyega Oralugbagbe

2025

Agri Horti Press

New Delhi-110 084

ISBN 978-93-83285-62-4 (Hardbound)

Published By

Agri Horti Press

New Delhi-110 084

E-mail: info@agrihorti.com

agrihortipress@gmail.com

Website: www.agrihorti.com

Printed at : **Replika Press Pvt. Ltd.**

Preface

Good nutrition is vital to good health. With such multiculturalism and modern society, food has established to be much more sophisticated and in variety. Our world is swept away with not we eat as a normal food but it has been triggering our taste buds by advertisements. Poor nutrition is the major problem in our country, unhealthy diet and physical inactivity, have produced to the growth of obesity in our country.

A strong and dynamic food of agricultural activities, improving value addition opportunities and creating surplus for export of agro food products. The science of food analysis has grown tremendously in its scope. New analysis techniques are being developed and existing techniques optimized.

To control food quality and acceptance within satisfactory limits, it is important to monitor the vital characteristics of raw products, ingredients and processed foods. This could be done by evaluating all foods or ingredients from a particular lot which is feasible if the analytical technique is rapid and non destructive. There is an increasing awareness of hazards posed to both human and animal health by mycotoxins in foodstuffs and feeds.

This book provides information on fruit and vegetable processing, milk and milk product processing, fish and marine product processing, cereal processing, determination of fat content, protein nitrogen, moisture content, vitamin C, sodium, potassium, flavour and carbohydrates. I hope that this book shall be ideal for undergraduate and postgraduate courses in food analysis of various disciplines including foods and nutrition, food science, food technology, soils agronomy, horticulture, agricultural sciences, dairy manufacturing and also an invaluable reference to professionals in the food industry.

All suggestions for future improvements for this book will be highly solicited.

Dr. Gboyega Oralugbagbe

Contents

Chapter 1

Fruit and Vegetable Processing

In developing countries agriculture is the mainstay of the economy. As such, it should be no surprise that agricultural industries and related activities can account for a considerable proportion of their output. Of the various types of activities that can be termed as agriculturally based, fruit and vegetable processing are among the most important.

Both established and planned fruit and vegetable processing projects aim at solving a very clearly identified development problem. This is that due to insufficient demand, weak infrastructure, poor transportation and perishable nature of the crops, the grower sustains substantial losses. During the post-harvest glut, the loss is considerable and often some of the produce has to be fed to animals or allowed to rot.

Even established fruit and vegetable canning factories or small/medium scale processing centres suffer huge loss due to erratic supplies. The grower may like to sell his produce in the open market directly to the consumer, or the produce may not be of high enough quality to process even though it might be good enough for the table. This means that processing capacities will be seriously underexploited.

The main objective of fruit and vegetable processing is to supply wholesome, safe, nutritious and acceptable food to consumers throughout the year.

Fruit and vegetable processing projects also aim to replace imported products like squash, yams, tomato sauces, pickles, etc., besides earning foreign exchange by exporting finished or semi-processed products.

The fruit and vegetable processing activities have been set up, or have to be established in developing countries for one or other of the following reasons:

- diversification of the economy, in order to reduce present dependence on one export commodity;

- government industrialization policy;
- reduction of imports and meeting export demands;
- stimulate agricultural production by obtaining marketable products;
- generate both rural and urban employment;
- reduce fruit and vegetable losses;
- improve farmers' nutrition by allowing them to consume their own processed fruit and vegetables during the off-season;
- generate new sources of income for farmers/artisans; and
- develop new value-added products.

What fruits and vegetables can be processed?

Practically any fruit and vegetable can be processed, but some important factors which determine whether it is worthwhile are:

a. the demand for a particular fruit or vegetable in the processed form;

b. the quality of the raw material, i.e. whether it can withstand processing; and

c. regular supplies of the raw material.

For example, a particular variety of fruit which may be excellent to eat fresh is not necessarily good for processing. Processing requires frequent handling, high temperature and pressure.

Many of the ordinary table varieties of tomatoes, for instance, are not suitable for making paste or other processed products. A particular mango or pineapple may be very tasty eaten fresh, but when it goes to the processing centre it may fail to stand up to the processing requirements due to variations in its quality, size, maturity, variety and so on.

Even when a variety can be processed, it is not suitable unless large and regular supplies are made available. An important processing centre or a factory cannot be planned just to rely on seasonal gluts; although it can take care of the gluts it will not run economically unless regular supplies are guaranteed.

To operate a fruit and vegetable processing centre efficiently it is of utmost importance to pre-organize growth, collection and transport of suitable raw material, either on the nucleus farm basis or using outgrowers.

Processing Planning

The secret of a well planned fruit and vegetable processing centre is that it must be designed to operate for as many months of the year as possible. This means the facilities, the buildings, the material handling and the equipment itself must be inter-linked and coordinated properly to allow as many products as possible to be handled at the same time, and yet the equipment must be versatile enough to be able to handle many products without major alterations.

A typical processing centre or factory should process four or five types of fruits harvested at different times of the year and two or three vegetables. This processing unit must also be capable of handling dried/dehydrated finished products, juices, pickles, tomato juice, ketchup and paste, jams, jellies and marmalades, semi-processed fruit products.

Advanced planning is necessary to process a large range of products in varied weather and temperature conditions, each requiring a special set of manufacturing and packaging formulae. The end result of the efforts should be a well-managed processing unit with lower initial investment.

A unit which is sensibly laid out and where one requirement co-relates to another, with a sound costing analysis, leads to an integrated operation.

Instead of over-sophisticated machinery, a sensible simple processing unit may be required when planned production is not very large and is geared mainly to meet the demand of the domestic market.

Location

The basic objective is to choose the location which minimizes the average production cost, including transport and handling.

It is an advantage, all other things being equal, to locate a processing unit near the fresh raw material supply. It is a necessity for proper handling of the perishable raw materials, it allows the processing unit to allow the product to reach its best stage of maturation and lessens injury from handling and deterioration from changes during long transportation after harvesting.

An adequate supply of good water, availability of manpower, proximity to rail or road transport facilities and adequate markets are other important requirements.

Processing Systems

a. Small-Scale Processing. This is done by small-scale farmers for personal subsistence or for sale in nearby markets. In this system, processing requires little investment: however, it is time consuming and tedious. Until recently, small-scale processing satisfied the needs of rural and urban populations. However, with the rising rates of population and urbanization growth and their more diversified food demands, there is need for more processed and diversified types of food.

b. Intermediate-Scale Processing. In this scale of processing, a group of small-scale processors pool their resources. This can also be done by individuals. Processing is based on the technology used by small-scale processors with differences in the type and capacity of equipment used. The raw materials are usually grown by the processors themselves or are purchased on contract from other farmers. These operations are usually located on the production site of in order to assure raw materials availability and reduce cost of transport. This system of processing can provide quantities of processed products to urban areas.

c. Large-Scale Processing. Processing in this system is highly mechanized and requires a substantial supply of raw materials for economical operation. This system requires a large capital investment and high technical and managerial skills. Because of the high demand for foods in recent years many large-scale factories were established in developing countries. Some succeeded, but the majority failed, especially in West Africa. Most of the failures were related to high labour inputs and relatively high cost, lack of managerial skills, high cost and supply instability of raw materials and changing governmental policies. Perhaps the most important reason for failure was lack of adequate quantity and regularity of raw material supply to factories. Despite the failure of these commercial operations, they should be able to succeed with better planning and management, along with the undertaking of more in-depth feasibility studies.

It can be concluded that all three types of processing systems have a place in developing countries to complement crop production to meet food demand. Historically, however, small and intermediate scale processing proved to be more successful than large-scale processing in developing countries.

Choice of Processing Technologies for Developing Countries

FAO maintains (in FAO, 1992c), that the basis for choosing a processing technology for developing countries ought to be to combine labour, material resources and capital so that not only the type and quantity of goods and services produced are taken into account, but also the distribution of their benefits and the prospects of overall growth. These should include:

a. increasing farmer/artisan income by the full utilization of available indigenous raw material and local manufacturing of part or all processing equipment;

b. cutting production costs by better utilization of local natural resources (solar energy) and reducing transport costs;

c. generating and distributing income by decentralizing processing activities and involving different beneficiaries in processing activities (investors, newly employed, farmers and small-scale industry);

d. maximizing national output by reducing capital expenditure and royalty payments, more effectively developing balance-of-payments deficits through minimizing imports (equipment, packing material, additives), and maximizing export-oriented production;

e. maximizing availability of consumer goods by maximization of high-quality, standard processed produce for internal and export markets, reducing post-harvest losses, giving added value to indigenous crops and increasing the volume and quality of agricultural output.

Knowledge and control of the means of production, local manufacturing of processing equipment and development of appropriate/new technologies and more suitable raw material for processing must all be better researched.

Decentralisation of activities must be maintained and coordinated. The introduction of more sophisticated processing equipment and packaging material must be subordinated to internal and export marketing references.

Choosing a technology solely to maximize profits can actually work against true development. Choice should also be based on a solid, long-term market opportunity to ensure viability.

The internal market should be given greater consideration, safeguarded and supported.

Training courses, at all levels, in processing and preservation of indigenous crops, must be expanded.

General Properties

Fruit and vegetables have many similarities with respect to their compositions, methods of cultivation and harvesting, storage properties and processing. In fact, many vegetables may be considered fruit in the true botanical sense. Botanically, fruits are those portions of the plant which house seeds. Therefore such items as tomatoes, cucumbers, eggplant, peppers, and others would be classified as fruits on this basis.

However, the important distinction between fruit and vegetables has come to be made on an usage basis. Those plant items that are generally eaten with the main course of a meal are considered to be vegetables. Those that are commonly eaten as dessert are considered fruits. That is the distinction made by the food processor, certain marketing laws and the consuming public, and this distinction will be followed in this document.

Vegetables are derived from various parts of plants and it is sometimes useful to associate different vegetables with the parts of the plant they represent since this provides clues to some of the characteristics we may expect in these items.

Fruit as a dessert item, is the mature ovaries of plants with their seeds. The edible portion of most fruit is the fleshy part of the pericarp or vessel surrounding the seeds. Fruit in general is acidic and sugary. They commonly are grouped into several major divisions, depending principally upon botanical structure, chemical composition and climatic requirements.

Berries are fruit which are generally small and quite fragile. Grapes are also physically fragile and grow in clusters. Melons, on the other hand, are large and have a tough outer rind. Drupes (stone fruit) contain single pits and include such items as apricots, cherries, peaches and plums. Pomes contain many pits, and are represented by apples, quince and pears.

Citrus fruit like oranges, grapefruit and lemons are high in citric acid. Tropical and subtropical fruits include bananas, dates, figs, pineapples, mangoes, and others which require warm climates, but exclude the separate group of citrus fruits.

Chemical Composition

Water

Vegetal cells contain important quantities of water. Water plays a vital role in the evolution and reproduction cycle and in physiological processes. It has effects on the storage period length and on the consumption of tissue reserve substances.

In vegetal cells, water is present in following forms:

- bound water or dilution water which is present in the cell and forms true solutions with mineral or organic substances;
- colloidal bound water which is present in the membrane, cytoplasm and nucleus and acts as a swelling agent for these colloidal structure substances; it is very difficult to remove during drying/dehydration processes;
- constitution water, directly bound on the chemical component molecules and which is also removed with difficulty.
- Vegetables contain generally 90-96% water while for fruit normal water content is between 80 and 90%.

Mineral Substances

Mineral substances are present as salts of organic or inorganic acids or as complex organic combinations (chlorophyll, lecithin, etc.); they are in many cases dissolved in cellular juice. Vegetables are more rich in mineral substances as compared with fruits. The mineral substance content is normally between 0.60 and 1.80% and more than 60 elements are present; the major elements are: K, Na, Ca, Mg, Fe, Mn, Al, P. Cl, S.

Among the vegetables which are especially rich in mineral substances are: spinach, carrots, cabbage and tomatoes. Mineral rich fruit includes: strawberries, cherries, peaches and raspberries. Important quantities of potassium (K) and absence of sodium chloride (NaCl) give a high dietetic value to fruit and to their processed products. Phosphorus is supplied mainly by vegetables.

Vegetables usually contain more calcium than fruit; green beans, cabbage, onions and beans contain more than 0.1% calcium. The calcium/phosphorus or Ca/P ratio is essential for calcium fixation in the human body; this value is considered normal at 0.7 for adults and at 1.0 for children. Some fruit are important for their Ca/P ratio above 1.0: pears, lemons, oranges and some temperate climate mountain fruits and wild berries. Even if its content in the human body is very low, iron (Fe) has an important role as a constituent of haemoglobin. Main iron sources are apples and spinach.Salts from fruit have a basic reaction; for this reason fruit consumption facilitates the neutralisation of noxious uric acid reactions and contributes to the acid-basic equilibrium in the blood.

Carbohydrates

Carbohydrates are the main component of fruit and vegetables and represent more than 90% of their dry matter. From energy point of view carbohydrates represent the most valuable of the food components; daily adult intake should contain about 500 g carbohydrates.

Carbohydrates play a major role in biological systems and in foods. They are produced by the process of photosynthesis in green plants. They may serve as structural components as in the case of cellulose; they may be stored as energy reserves as in the case of starch in plants; they may function as essential components of nucleic acids as in the case of ribose; and as components of vitamins such as ribose and riboflavin.

Carbohydrates can be oxidized to furnish energy, and glucose in the blood is a ready source of energy for the human body. Fermentation of carbohydrates by yeast and other microorganisms can yield carbon dioxide, alcohol, organic acids and other compounds.

Some properties of sugars: Sugars such as glucose, fructose, maltose and sucrose all share the following characteristics in varying degrees, related to fruit and vegetable technology:

- they supply energy for nutrition;
- they are readily fermented by micro-organisms;
- in high concentrations they prevent the growth of micro-organisms, so they may be used as a preservative;
- on heating they darken in colour or caramelise;
- some of them combine with proteins to give dark colours known as the browning reaction.

Some properties of starches

- They provide a reserve energy source in plants and supply energy in nutrition;
- they occur in seeds and tubers as characteristic starch granules.

Some properties of celluloses and hemicelluloses

- They are abundant in the plant kingdom and act primarily as supporting structures in the plant tissues;
- they are insoluble in cold and hot water;
- they are not digested by man and so do not yield energy for nutrition;
- the fibre in food which produces necessary roughage is largely cellulose.

Some properties of pectins and carbohydrate gums

- Pectins are common in fruits and vegetables and are gum-like (they are found in and between cell walls) and help hold the plant cells together;
- pectins in colloidal solution contribute to viscosity of the tomato paste;
- pectins in solution form gels when sugar and acid are added; this is the basis of jelly manufacture.

Fats

Generally fruit and vegetables contain very low level of fats, below 0.5%. However, significant quantities are found in nuts (55%), apricot kernel (40%), grapes seeds (16%), apple seeds (20%) and tomato seeds (18%).

Organic Acids

Fruit contains natural acids, such as citric acid in oranges and lemons, malic acid of apples, and tartaric acid of grapes. These acids give the fruits tartness and slow down bacterial spoilage.

We deliberately ferment some foods with desirable bacteria to produce acids and this give the food flavour and keeping quality. Examples are fermentation of cabbage to produce lactic acid and yield sauerkraut and fermentation of apple juice to produce first alcohol and then acetic acid to obtain vinegar.

Organic acids influence the colour of foods since many plant pigments are natural pH indicators.

With respect to bacterial spoilage, a most important contribution of organic acids is in lowering a food's pH. Under anaerobic conditions and slightly above a pH of 4.6, Clostridium botulinum can grow and produce lethal toxins. This hazard is absent from foods high in organic acids resulting in a pH of 4.6 and less.

Acidity and sugars are two main elements which determine the taste of fruit. The sugar/acid ratio is very often used in order to give a technological characterisation of fruits and of some vegetables.

Nitrogen-containing Substances

These substances are found in plants as different combinations: proteins, amino acids, amides, amines, nitrates, etc. Vegetables contain between 1.0 and 5.5 % while in fruit nitrogen-containing substances are less than 1% in most cases.

Among nitrogen containing substances the most important are proteins; they have a colloidal structure and, by heating, their water solution above 50°C an one-way reaction makes them insoluble. This behaviour has to be taken into account in heat processing of fruits and vegetables.

From a biological point of view vegetal proteins are less valuable then animal ones because in their composition all essential amino-acids are not present.

Vitamins

Vitamins are defined as organic materials which must be supplied to the human body in small amounts apart from the essential amino-acids or fatty acids.

Vitamins function as enzyme systems which facilitate the metabolism of proteins, carbohydrates and fats but there is growing evidence that their roles in maintaining health may extend yet further.

The vitamins are conveniently divided into two major groups, those that are fat-soluble and those that are water-soluble. Fat-soluble vitamins are A, D, E and K. Their absorption by the body depends upon the normal absorption of fat from the diet. Water-soluble vitamins include vitamin C and several members of the vitamin B complex.

Enzymes

Enzymes are biological catalysts that promote most of the biochemical reactions which occur in vegetable cells.

Some properties of enzymes important in fruit and vegetable technology are the following:

- in living fruit and vegetables enzymes control the reactions associated with ripening;
- after harvest, unless destroyed by heat, chemicals or some other means, enzymes continue the ripening process, in many cases to the point of spoilage - such as soft melons or overripe bananas;
- because enzymes enter into a vast number of biochemical reactions in fruits and vegetable, they may be responsible for changes in flavour, colour, texture and nutritional properties;
- the heating processes in fruit and vegetables manufacturing/processing are designed not only to destroy micro-organisms but also to deactivate enzymes and so improve the fruit and vegetables' storage stability.

Enzymes have an optimal temperature - around +50°C where their activity is at maximum. Heating beyond this optimal temperature deactivates the enzyme. Activity of each enzyme is also characterised by an optimal pH.

In fruit and vegetable storage and processing the most important roles are played by the enzymes classes of hydrolases (lipase, invertase, tannase, chlorophylase, amylase, cellulase) and oxidoreductases (peroxidase, tyrosinase, catalase, ascorbinase, polyphenoloxidase).

Sources of colour and colour changes

In addition to a great range of textures, much of the interest that fruits and vegetables add to our diets is due to their delightful and variable colours. The pigments and colour precursors of fruit and vegetables occur for the most part in the cellular plastic inclusions such as the chloroplasts and other chromoplasts, and to a lesser extent dissolved in fat droplets or water within the cell protoplast and vacuoles.

These pigments are classified into four major groups which include the chlorophylls, carotenoids, anthocyanins, and anthoanthins. Pigments belonging to the latter two groups also are referred to as flavonoids, and include the tannins.

The Chlorophylls: The chlorophylls are contained mainly within the chloroplasts and have a primary role in the photosynthetic production of carbohydrates from carbon dioxide and water. The bright green colour of leaves and other parts of plants is largely due to the oilsoluble chlorophylls, which in nature are bound to protein molecules in highly organised complexes.

When the plant cells are killed by ageing, processing, or cooking, the protein of these complexes is denatured and the chlorophyll may be released. Such chlorophyll is highly unstable and rapidly changes in colour to olive green and brown. This colour change is believed to be due to the conversion of chlorophyll to the compound pheophytin.

Conversion to pheophytin is favoured by acid pH but does not occur readily under alkaline conditions. For this reason peas, beans, spinach, and other green vegetables which tend to lose their bright green colours on heating can be largely protected against such colour changes by the addition of sodium bicarbonate or other alkali to the cooking or canning water.

However, this practice is not looked upon favourably nor used commercially because alkaline pH also has a softening effect on cellulose and vegetable texture and also destroys vitamin C and thiamin at cooking temperatures.

The Carotenoids: Pigments belonging to this group are fat-soluble and range in colour from yellow through orange to red. They often occur along with the chlorophylls in the chloroplasts, but also are present in other chromoplasts and may occur free in fat droplets. Important carotenoids include the orange carotenes of carrot, maize, apricot, peach, citrus fruits, and squash; the red lycopene of tomato, watermelon, and apricot; the yellow-orange xanthophyll of maize, peach, paprika and squash; and the yellow-orange crocetin of the spice saffron. These and other carotenoids seldom occur singly within plant cells.

A major importance of some of the carotenoids is their relationship to vitamin A. A molecule of orange beta-carotene is converted into two molecules of colourless vitamin A in the animal body. Other carotenoids like alpha-carotene, gamma-carotene, and cryptoxanthin also are precursors of vitamin A, but because of minor differences in chemical structure one molecule of each of these yields only one molecule of vitamin A.

In food processing the carotenoids are fairly resistant to heat, changes in pH, and water leaching since they are fat-soluble. However, they are very sensitive to oxidation, which results in both colour loss and destruction of vitamin A activity.

The Flavonoids: Pigments and colour precursors belonging to this class are water-soluble and commonly are present in the juices of fruit and vegetables. The flavonoids include the purple, blue, and red anthocyanins of grapes, berries, plump, eggplant, and cherry; the yellow anthoxanthins of light coloured fruit and vegetables such as apple, onion, potato, and cauliflower, and the colourless catechins and leucoanthocyanins which are food tannins and are found in apples, grapes, tea, and other plant tissues. These colourless tannin compounds are easily converted to brown pigments upon reaction with metal ions.

Properties of the anthocyanins include a shifting of colours with pH. Thus many of the anthocyanins which are violet or blue in alkaline media become red upon addition of acid.

Cooking of beets with vinegar tends to shift the colour from a purplish red to a brighter red, while alkaline water can influence the colour of red fruits and vegetables toward violet and gray-blue.

The anthocyanins also tend toward the violet and blue hues upon reaction with metal ions, which is one reason for lacquering the inside of metal cans when the true colour of anthocyanin-containing fruits and vegetables is to be preserved.

The water-soluble property of anthocyanins also results in easy leaching of these pigments from cut fruit and vegetables during processing and cooking.

The yellow anthoxanthins also are pH sensitive tending toward a deeper yellow in alkaline media. Thus potatoes or apples become somewhat yellow when cooked in water with a pH of 8 or higher, which is common in many areas. Acidification of the water to pH 6 or lower favours a whiter colour.

The colourless tannin compounds upon reaction with metal ions form a range of dark coloured complexes which may be red, brown, green, grey, or black. The various shades of these coloured complexes depend upon the particular tannin, the specific metal ion, pH, concentration of the complex, and other factors not yet fully understood.

Water-soluble tannins appear in the juices squeezed from grapes, apples, and other fruits as well as the brews from extraction of tea and coffee. The colour and clarity of tea are influenced by the hardness and pH of the brewing water. Alkaline waters that contain calcium and magnesium favour the formation of dark brown tannin complexes which precipitate when the tea is cooled.

If acid in the form of lemon juice is added to such tea its colour lightens and the precipitate tends to dissolve. Iron from equipment or from pitted tin cans has caused a number of unexpected colours to develop in products containing tannins, such as coffee, cocoa and foods flavoured with these.

The tannins are also important because they have an astringency which influences flavour and contributes body to such beverages as tea, wine, apple cider, etc.

Stability of Nutrients

One of the principal responsibilities of the food scientist and food technologist is to preserve food nutrients through all phases of food acquisition, processing, storage, and preparation.

This shows the stability of vitamins, essential amino acids, and minerals to acid, air, light, and heat, and gives an indication of possible cooking losses. Vitamin A is highly sensitive to acid, air, light and heat; vitamin C to alkalinity, air, light and heat; vitamin D to alkalinity, air, light and heat; thiamin to alkalinity, air, and heat in alkaline solutions; etc. Cooking losses of some essential nutrients may be in excess of 75%. In modern food processing operations, however, losses are seldom in excess of 25% .

The ultimate nutritive value of a food results from the sum total of losses incurred throughout its history - from farmer to consumer. Nutrient value begins with genetics of the plant and animal. The farmland fertilization program affects tissue composition of plants, and animals consuming these plants. The weather and degree of maturity at harvest affect tissue composition.

Storage conditions before processing affect vitamins and other nutrients. Washing, trimming, and heat treatments affect nutrient content. Canning, evaporating, drying, and freezing alter nutritional values, and the choices of times and temperatures in these operations frequently must be balanced between good bacterial destruction and minimum nutrient destruction.

Packaging and subsequent storage affect nutrients. One of the most important factors is the final preparation of the food in the home and the restaurant - the steam table can destroy much of what has been preserved through all prior manipulations.

Deterioration Factors and their Control

A summary of overall deterioration reactions in fruits and vegetables is presented below.

Enzymic Changes

Enzymes which are endogenous to plant tissues can have undesirable or desirable consequences. Examples involving endogenous enzymes include a) the post-harvest senescence and spoilage of fruit and vegetables; b) oxidation of phenolic substances in plant tissues by phenolase (leading to browning); c) sugar - starch conversion in plant tissues by amylases; d) post-harvest demethylation of pectic substances in plant tissues (leading to softening of plant tissues during ripening, and firming of plant tissues during processing).

The major factors useful in controlling enzyme activity are: temperature, water activity, pH, chemicals which can inhibit enzyme action, alteration of substrates, alteration of products and pre-processing control.

Chemical Changes

The two major chemical changes which occur during the processing and storage of foods and lead to a deterioration in sensory quality are lipid oxidation and non-enzymatic browning. Chemical reactions are also responsible for changes in the colour and flavour of foods during processing and storage.

Lipid oxidation rate and course of reaction is influenced by light, local oxygen concentration, high temperature, the presence of catalysts (generally transition metals such as iron and copper) and water activity. Control of these factors can significantly reduce the extent of lipid oxidation in foods.

Non-enzymic browning is one of the major causes of deterioration which occurs during storage of dried and concentrated foods. The non-enzymic browning, or Maillard reaction, can be divided into three stages: a) early Maillard reactions which are chemically well-defined steps without browning; b) advanced Maillard reactions which lead to the formation of volatile or soluble substances; and c) final Maillard reactions leading to insoluble brown polymers.

Chlorophylls. Almost any type of food processing or storage causes some deterioration of the chlorophyll pigments. Phenophytinisation (with consequent formation of a dull olivebrown phenophytin) is the major change; this reaction is accelerated by heat and is acid catalysed.

Other reactions are also possible. For example, dehydrated products such as green peas and beans packed in clear glass containers undergo photo-oxidation and loss of desirable colour.

Anthocyanins. These are a group of more than 150 reddish water-soluble pigments that are very widespread in the plant kingdom. The rate of anthocyanin destruction is pH dependent, being greater at higher pH values. Of interest from a packaging point of view is the ability of some anthocyanins to form complexes with metals such as Al, Fe, Cu and Sn.

These complexes generally result in a change in the colour of the pigment (for example, red sour cherries react with tin to form a purple complex) and are therefore undesirable. Since metal packaging materials such as cans could be sources of these metals, they are usually coated with special organic linings to avoid these undesirable reactions.

Carotenoids. The carotenoids are a group of mainly lipid soluble compounds responsible for many of the yellow and red colours of plant and animal products. The main cause of carotenoid degradation in foods is oxidation. The mechanism of oxidation in processed foods is complex and depends on many factors. The pigments may auto-oxidise by reaction with atmospheric oxygen at

In fruit and vegetables, enzymically generated compounds derived from long-chain fatty acids play an extremely important role in the formation of characteristic flavours. In addition, these types of reactions can lead to significant off-flavours. Enzyme-induced oxidative breakdown of unsaturated fatty acids occurs extensively in plant tissues and this yield characteristic aromas associated with some ripening fruits and disrupted tissues.

The permeability of packaging materials is of importance in retaining desirable volatile components within packages, or in permitting undesirable components to permeate through the package from the ambient atmosphere.

Nutritional Quality

The four major factors which affect nutrient degradation and can be controlled to varying extents by packaging are light, oxygen concentration, temperature and water activity. However, because of the diverse nature of the various nutrients as well as the chemical heterogeneity within each class of compounds and the complex interactions of the above variables, generalizations about nutrient degradation in foods will inevitably be broad ones.

Vitamins. Ascorbic acid is the most sensitive vitamin in foods, its stability varying markedly as a function of environmental conditions such as pH and the concentration of trace metal ions and oxygen. The nature of the packaging material can significantly affect the stability of ascorbic acid in foods. The effectiveness of the material as a barrier to moisture and oxygen as well as the chemical nature of the surface exposed to the food are important factors.

For example, problems of ascorbic acid instability in aseptically packaged fruit juices have been encountered because of oxygen permeability of the package and the oxygen dependence of the ascorbic acid degradation reaction.

Also, because of the preferential oxidation of metallic tin, citrus juices packaged in cans with a tin contact surface exhibit greater stability of ascorbic acid than those in enamelled cans or glass containers. The aerobic and anaerobic degradation reactions of ascorbic acid in reduced-moisture foods have been shown to be highly sensitive to water activity, the reaction rate increasing in an exponential fashion over the water activity range of 0.1-0.8.

Physical Changes

One major undesirable physical change in food powders is the absorption of moisture as a consequence of an inadequate barrier provided by the package; this results in caking. It can occur either as a result of a poor selection of packaging material in the first place, or failure of the package integrity during storage. In general, moisture absorption is associated with increased cohesiveness.

Anti-caking agents are very fine powders of an inert chemical substance that are added to powders with much larger particle size in order to inhibit caking and improve flowability. Studies in onion powders showed that at ambient temperature, caking does not occur at water activities of less than about 0.4.

At higher activities, however, (aw > 0.45) the observed time to caking is inversely proportional to water activity, and at these levels anti-caking agents are completely ineffective. It appears that while they reduce inter-particle attraction and interfere with the continuity of liquid bridges, they are unable to cover moisture sorption sites.

Biological Changes

Microbiological

Micro-organisms can make both desirable and undesirable changes to the quality of foods depending on whether or not they are introduced as an essential part of the food preservation process or arise unintentionally and subsequently grow to produce food spoilage.

The two major groups of micro-organisms found in foods are bacteria and fungi, the latter consisting of yeasts and moulds. Bacteria are generally the fastest growing, so that in conditions favourable to both, bacteria will usually outgrow fungi.

Foods are frequently classified on the basis of their stability as non-perishable, semi-perishable and perishable. For example, hermetically sealed and heat processed (e.g. canned) foods are generally regarded as non-perishable. However, they may become perishable under certain circumstances when an opportunity for recontamination is afforded following processing.

Such an opportunity may arise if the can seams are faulty, or if there is excessive corrosion resulting in internal gas formation and eventual bursting of the can. Spoilage may also take place when the canned food is stored at unusually high temperatures: thermophilic spore-forming bacteria may multiply, causing undesirable changes such as flat sour spoilage.

Low moisture content foods such as dried fruit and vegetables are classified as semi-perishable. Frozen foods, though basically perishable, may be classified as semi-perishable provided that they are properly stored at freezer temperatures.

The majority of foods (e.g. meat and fish, milk, eggs and most fresh fruits and vegetables) are classified as perishable unless they have been processed in some way. Often, the only form of processing which such foods receive is to be packaged and kept under controlled temperature conditions.

The species of micro-organisms which cause the spoilage of particular foods are influenced by two factors: a) the nature of the foods and b) their surroundings. These factors are referred to as intrinsic and extrinsic parameters.

The intrinsic parameters are an inherent part of the food: pH, a_w, nutrient content, antimicrobial constituents and biological structures. The extrinsic parameters of foods are those properties of the storage environment that affect both the foods and their microorganisms. The growth rate of the micro-organisms responsible for spoilage primarily depends on these extrinsic parameters: temperature, relative humidity and gas compositions of the surrounding atmosphere.

The protection of packaged food from contamination or attack by micro-organisms depends on the mechanical integrity of the package (*e.g.* the absence of breaks and seal imperfections), and on the resistance of the package to penetration by micro-organisms.

Metal cans which are retorted after filling can leak during cooling, admitting any microorganisms which may be present in the cooling water, even when the double seam is of a high quality. This fact is widely known in the canning industry and is the reason for the mandatory chlorination of cannery cooling water.

Extensive studies on a variety of plastic films and metal foils have shown that microorganisms (including mounds, yeasts and bacteria) cannot penetrate these materials in the absence of pinholes.

In practice, however, thin sheets of packaging materials such as aluminium and plastic do contain pinholes. There are several safeguards against the passage of micro-organisms through pinholes in films:

- because of surface tension effects, micro-organisms cannot pass through very small pinholes unless the micro-organisms are suspended in solutions containing wetting agents and the pressure outside the package is greater than that within;
- materials of packaging are generally used in thicknesses such that pinholes are very infrequent and small;
- for applications in which package integrity is essential (such as sterilisation of food in pouches), adequate test methods are available to assure freedom from bacterial recontamination.

Macrobiological

Insect pests

Warm humid environments promote insect growth, although most insects will not breed if the temperature exceeds about 35 °C or falls below 10 °C. Also many insects cannot reproduce satisfactorily unless the moisture content of their food is greater than about 11%.

The main categories of foods subject to pest attack are cereal grains and products derived from cereal grains, other seeds used as food (especially legumes), dairy products such as cheese and milk powders, dried fruits, dried and smoked meats and nuts.

As well as their possible health significance, the presence of insects and insect excrete in packaged foods may render products unsaleable, causing considerable economic loss, as well as reduction in nutritional quality, production of off-flavours and acceleration of decay processes due to creation of higher temperatures and moisture levels.

Early stages of infestation are often difficult to detect; however, infestation can generally be spotted not only by the presence of the insects themselves but also by the products of their activities such as webbing, clumped-together food particles and holes in packaging materials.

Unless plastic films are laminated with foil or paper, insects are able to penetrate most of them quite easily, the rate of penetration usually being directly related to film thickness. In general, thicker films are more resistant than thinner films, and oriented films tend to be more effective than cast films. The looseness of the film has also been reported to be an important factor, loose films being more easily penetrated than tightly fitted films.

Generally, the penetration varies depending on the basic resin from which the film is made, on the combination of materials, on the package structure, and of the species and stage of insects involved. The relative resistance to insect penetration of some flexible packaging materials is as follows:

- excellent resistance: polycarbonate; poly-ethylene-terephthalate;
- good resistance: cellulose acetate; polyamide; polyethylene (0.254 mm); polypropylene (biaxially oriented); poly-vinyl-chloride (unplasticised);
- fair resistance: acrylonitrile; poly-tetra-fluoro-ethylene; polyethylene (0.123 mm);
- poor resistance: regenerated cellulose; corrugated paper board; kraft paper; polyethylene (0.0254 - 0.100 mm); paper/foil/polyethylene laminate pouch; poly-vinylchloride (plasticised).

Some simple methods for obtaining insect resistance of packaging materials are as following:

- select a film and a film thickness that are inherently resistant to insect penetration;
- use shrink film over-wraps to provide an additional barrier;
- seal carton flaps completely.

Rodents

Rats and mice carry disease-producing organisms on their feet and/or in their intestinal tracts and are known to harbour salmonella of serotypes frequently associated with food-borne infections in humans. In addition to the public health consequences of rodent populations in close proximity to humans, these animals also compete intensively with humans for food.

Rats and mice gnaw to reach sources of food and drink and to keep their teeth short. Their incisor teeth are so strong that rats have been known to gnaw through lead pipes and unhardened concrete, as well as sacks, wood and flexible packaging materials. Proper sanitation in food processing and storage areas is the most effective weapon in the fight against rodents, since all packaging materials apart from metal and glass containers can be attacked by rats and mice.

Methods of Reducing Deterioration

Knowledge of deterioration factors and the way they act, including the rates of deterioration to a specific category of food, means that it is possible to list the ways of lowering or stopping the action and obtaining fruit and vegetable preservation.

In order to maintain their nutritional value and organoleptic properties and because of technical-economical considerations, not all the identified means against deterioration actually have practical applications for fruit and vegetable preservation.

These technical means can be summarized as follows:

Table 1.1 Technical Methods of Reducing Food Deterioration

Physical	Heating
	Cooling
	Lowering of water content Drying/dehydration. Concentration
	Sterilising filtration
	Irradiation
	Other physical means (high pressure, vacuum, inert gases)
Chemical	Salting
	Smoking
	Sugar addition
	Artificial acidification
	Ethyl alcohol addition
	Antiseptic substance action
Biochemical	Lactic fermentation (natural acidification)
	Alcoholic fermentation

This classification of methods of reducing deterioration presents some difficulties because their preservation effects are physical, physico-chemical, chemical and biochemical complex phenomena which rarely act in isolation. Normally they take place together or one after the other.

From the whole list of possible methods of reducing deterioration, over the years, some procedures for fruit and vegetable preservation have found practical application.

Table 1.2 Procedures for Fruit and Vegetable Preservation

Procedures	Practical applications
Fresh storage	Fruits, vegetables
Cold storage	Fruits, vegetables
Freezing	Fruits, vegetables
Drying/dehydration	Fruits, vegetables
Concentration	Fruit and vegetable juices
Chemical preservation	Fruit semi-processed
Preservation with sugar	Fruit products/preserves
Pasteurization	Fruit and vegetable juices
Sterilisation	Fruits, vegetables
Sterilising filtration	Fruit juices
Irradiation	Fruits, vegetables

These preservation procedures have two main characteristics as far as being applied to all food products is concerned:

- some of them are applied only to one or some categories of foods; others can be used across the board and thus a wider application (cold storage, freezing, drying/dehydration, sterilisation, etc.);
- some guarantee food preservation on their own while others require combination with other procedures, either as principal or as auxiliary processes in order to assure preservation (for example smoking has to be preceded by salting).

Combined Preservation Procedures

In practice preservation procedures aim at avoiding microbiological and biochemical deterioration which are the principal forms of deterioration. Even with all recent progress achieved in this field, no single one of these technological procedures applied alone can be considered wholly satisfactory from a microbiological, physico-chemical and organoleptic point of view, even if to a great extent the food value is assured.

Thus, heat sterilisation cannot be applied in order to destroy all micro-organisms present in foods without inducing non desirable modifications. Preservation by dehydration/drying assures microbiological stability but has the drawback of undesirable modifications that appear during storage: vitamin losses, oxidation phenomena, etc.

Starting with these considerations, the actual tendency in food preservation is to study the application of combined preservation procedures, aiming at the realisation of maximum efficiency from a microbiological and biological point of view, with reduction to a minimum of organoleptical degradation and decrease in food value.

The principles of combined preservation procedures are:

- avoid or reduce secondary (undesirable) effects in efficient procedures for microbiological preservation;
- avoid qualitative degradation appearing during storage of products preserved by efficient procedures from a microbiological point of view;
- increase microbiological efficiency of preservation procedures by supplementary means;
- combine preservation procedures in order to obtain maximum efficiency from a microbiological point of view, by specific action on various types of micro-organisms present;
- establish combined factors that act simultaneously on bacterial cells.

Research and applications in this direction were followed by microbiological and biochemical way, obtaining a serial of combination of preservation procedures with the possibility of application in industrial practice.

Fresh fruit and vegetable storage can be combined with:

- storage in controlled atmosphere where carbon dioxide and oxygen levels are monitored, increasing concentration of CO2 and lowering that of oxygen according to fruit species. Excellent results were obtained for pomace fruit; in particular the storage period for apples has been extended. Application of this combined procedure requires airtight storage rooms.
- storage in an environment containing ethylene oxide; this accelerates ripening in some fruit: tomatoes, bananas, mangoes, etc.

Cold storage can be combined with storage in an environment with added of carbon dioxide, sulphur dioxide, etc. according to the nature of product to be preserved.

Preservation by drying/dehydration can be combined with:

- freezing: fresh fruit and vegetables are dehydrated up to the point where their weight is reduced by 50% and then they are preserved by freezing.

This procedure (freeze-drying) combines the advantages of drying (reduction of volume and weight) with those of freezing (maintaining vitamins and to a large extent organoleptic properties).

A significant advantage of this process is the short drying time in so far as it is not necessary to go beyond the inflexion point of the drying curve. The finished products after defreezing and rehydration/reconstitution are of a better quality compared with products obtained by dehydration alone.

- cold storage of dried/dehydrated vegetables in order to maintain vitamin C; storage temperature can be varied with storage time and can be at -8° C for a storage time of more than one year, with a relative humidity of 70-75 %.

- packaging under vacuum or in inert gases in order to avoid action of atmospheric oxygen;, mainly for products containing beta-carotene.
- chemical preservation: a process used intensively for prunes and which has commercial applications is to rehydrate the dried product up to 35 % using a bath containing hot 2 % potassium sorbate solution. Another possible application of this combined procedure is the initial dehydration up to 35% moisture followed by immersion in same bath as explained above; this has the advantage of reducing drying time and producing minimum qualitative degradation. Both applications suppress the dehydrated products reconstitution (rehydration) step before consumption.
- packaging in the presence of desiccants (calcium oxide, anhydrous calcium chloride, etc.) in order to reduce water vapour content in the package, especially for powdered products.

Preservation by concentration, carried out by evaporation, is combined with cold storage during warm season for tomato paste (when water content cannot be reduced under the limit needed to inhibit moulds and yeasts, e.g. $a_w = 0.70...0.75$).

Chemical preservation is combined with:

- acidification of food medium (lowering pH);
- using combined chemical preservatives.

Preservation by lactic fermentation (natural acidification) can be combined with cold storage for pickles in order to prolong storage time or shelf-life.

Preservation with sugar is combined with pasteurization for some preserves having a sugar content below 65%.

General Procedures for Fruit and Vegetable Preservation

Fresh Storage

Fresh fruit and vegetable storage

Once fruit is harvested, any natural resistance to the action of spoiling micro-organisms is lost. Changes in enzymatic systems of the fruit also occur on harvest which may also accelerate the activity of spoilage organisms.

Means that are commonly used to prevent spoilage of fruits must include:

- care to prevent cutting or bruising of the fruit during picking or handling;
- refrigeration to minimise growth of micro-organisms and reduce enzyme activity;
- packaging or storage to control respiration rate and ripening;
- use of preservatives to kill micro-organisms on the fruit.

A principal economic loss occurring during transportation and/or storage of produce such as fresh fruit is the degradation which occurs between the field and the ultimate destination due to the effect of respiration. Methods to reduce such degradation are as follows:

- refrigerate the produce to reduce the rate of respiration;
- vacuum cooling;
- reduce the oxygen content of the environment in which the produce is kept to a value not above 5% of the atmosphere but above the value at which anaerobic respiration would begin. When the oxygen concentration is reduced within 60 minutes the deterioration is in practice negligible.

The following is a summary of some recent developments in post-harvest technology of fresh fruits and vegetables (Source: Thompson, 1989).

Harvest maturity. This is particularly important with fruit for export. One recent innovation is the measurement of resonant frequency of the fruit which should enable the grading out of over mature and under-mature fruit before they are packed for export.

Harvest method. Considerable research is continuing on mechanical harvesting of perishable crops with a view to minimising damage. In fruit trees, controlling their height by use of dwarfing rootstocks, pruning and growth regulating chemicals will lead to easier, cheaper more accurate harvesting.

Handling systems. Field packing of various vegetables for export has been carried out for many years. In the last decade or so this has been applied, in selected cases, to a few tropical fruit types. Where this system can be practiced it has considerable economic advantages in saving the cost of building, labour and equipment and can result in lower levels of damage into crops.

Pre-cooling. Little innovation has occurred in crop pre-cooling over the last decade. However high velocity, high humidity forced air systems have continued to be developed and refined. These are suitable for all types of produce and are relatively simple to build and operate and, while not providing the speed of cooling of a vacuum or hydrocooler, have the flexibility to be used with almost all crops.

Chemicals. There is a very strong health lobby whose objective is to reduce the use of chemicals in agriculture and particularly during the post harvest period. Every year sees the prohibition of the use of commonly used post-harvest chemicals. New ways need to be developed to control post-harvest diseases, pest and sprouting.

Coatings. Slowing down the metabolism of fruit and vegetables by coating them with a material which affects their gaseous exchange is being tested and used commercially on a number of products.

Controlled environment transport. Recent innovations in this technique have produced great progress as a result of the development and miniaturisation of equipment to measure carbon dioxide and oxygen. Several companies now offer containers where the levels of these two gases can be controlled very precisely.

Preservation by Drying/Dehydration

The technique of drying is probably the oldest method of food preservation practiced by mankind. The removal of moisture prevents the growth and reproduction of micro-organisms causing decay and minimises many of the moisture mediated deterioration reactions.

It brings about substantial reduction in weight and volume minimising packing, storage and transportation costs and enable storability of the product under ambient temperatures, features especially important for developing countries. The sharp rise in energy costs has promoted a dramatic upsurge in interest in drying world-wide over the last decade.

Heat and mass transfer

Dehydration involves the application of heat to vaporise water and some means of removing water vapour after its separation from the fruit/vegetable tissues. Hence it is a combined/simultaneous (heat and mass) transfer operation for which energy must be supplied.

A current of air is the most common medium for transferring heat to a drying tissue and convection is mainly involved.

The two important aspects of mass transfer are:

- the transfer of water to the surface of material being dried and
- the removal of water vapour from the surface.

In order to assure products of high quality at a reasonable cost, dehydration must occur fairly rapidly. Four main factors affect the rate and total drying time:

- the properties of the products, especially particle size and geometry;
- the geometrical arrangement of the products in relation to heat transfer medium (drying air);
- the physical properties of drying medium/ environment;
- the characteristics of the drying equipment.

It is generally observed with many products that the initial rate of drying is constant and then decreases, sometimes at two different rates. The drying curve is divided into the constant rate period and the falling rate period.

Surface area. Generally the fruit and vegetables to be dehydrated are cut into small pieces or thin layers to speed heat and mass transfer. Subdivision speeds drying for two reasons:

- large surface areas provide more surface in contact with the heating medium (air) and more surface from which moisture can escape;
- smaller particles or thinner layers reduce the distance heat must travel to the centre of the food and reduce the distance through which moisture in the centre of the food must travel to reach the surface and escape.

Temperature. The greater the temperature difference between the heating medium and the food the greater will be the rate of heat transfer into the food, which provides the driving force for moisture removal. When the heating medium is air, temperature plays a second important role.

As water is driven from the food in the form of water vapour it must be carried away, or else the moisture will create a saturated atmosphere at the food's surface which will slow down the rate of subsequent water removal. The hotter the air the more moisture it will hold before becoming saturated.

Thus, high temperature air in the vicinity of the dehydrating food will take up the moisture being driven from the food to a greater extent than will cooler air. Obviously, a greater volume of air also can take up more moisture than a lesser volume of air.

Air velocity. Not only will heated air take up more moisture than cool air, but air in motion will be still more effective. Air in motion, that is, high velocity air, in addition to taking up moisture will sweep it away from the drying food's surface, preventing the moisture from creating a saturated atmosphere which would slow down subsequent moisture removal. This is why clothes dry more rapidly on a windy day.

Some other phenomena influence the drying process and a few elements are summarised below.

Dryness of air. When air is the drying medium of food, the drier the air the more rapid is the rate of drying. Dry air is capable of absorbing and holding moisture. Moist air is closer to saturation and so can absorb and hold less additional moisture than if it were dry. But the dryness of the air also determines how low a moisture content the food product can be dried to.

Atmospheric pressure and vacuum. If food is placed in a heated vacuum chamber the moisture can be removed from the food at a lower temperature than without a vacuum. Alternatively, for a given temperature, with or without vacuum, the rate of water removal from the food will be greater in the vacuum. Lower drying temperatures and shorter drying times are especially important in the case of heat-sensitive foods.

Evaporation and temperature. As water evaporates from a surface it cools the surface. The cooling is largely the result of absorption by the water of the latent heat of phase change from liquid to gas.

In doing this the heat is taken from the drying air or the heating surface and from the hot food, and so the food piece or droplet is cooled.

Time and temperature. Since all important methods of food dehydration employ heat, and food constituents are sensitive to heat, compromises must be made between maximum possible drying rate and maintenance of food quality.

As is the case in the use of heat for pasteurization and sterilisation, with few exceptions drying processes which employ high temperatures for short times do less damage to food than drying processes employing lower temperatures for longer times.

Thus, vegetable pieces dried in a properly designed oven in four hours would retain greater quality than the same products sun dried over two days.

Several drying processes will achieve dehydration in a matter of minutes or even less if the food is sufficiently subdivided.

Drying techniques

Several types of dryers and drying methods, each better suited for a particular situation, are commercially used to remove moisture from a wide variety of food products including fruit and vegetables.

While sun drying of fruit crops is still practiced for certain fruit such as prunes, figs, apricots, grapes and dates, atmospheric dehydration processes are used for apples, prunes, and several vegetables; continuous processes as tunnel, belt trough, fluidised bed and foam-mat drying are mainly used for vegetables.

Spray drying is suitable for fruit juice concentrates and vacuum dehydration processes are useful for low moisture / high sugar fruits like peaches, pears and apricots.

Factors on which the selection of a particular dryer/ drying method depends include:

- form of raw material and its properties;
- desired physical form and characteristics of dried product;
- necessary operating conditions;
- operation costs.

There are three basic types of drying process:

- sun drying and solar drying;
- atmospheric drying including batch (kiln, tower and cabinet dryers) and continuous (tunnel, belt, belt-trough, fluidised bed, explosion puff, foam-mat, spray, drum and microwave);
- sub-atmospheric dehydration (vacuum shelf/belt/drum and freeze dryers).

The scope has been expanded to include use of low temperature, low energy process like osmotic dehydration.

Preservation by Concentration

Foods are concentrated for many of the same reasons that they are dehydrated; concentration can be a form of preservation but this is true only for some foods. Concentration reduces weight and volume and results in immediate economic advantages.

Nearly all liquid foods which are dehydrated are concentrated before they are dried. This is because in the early stages of water removal, moisture can be more economically removed in highly efficient evaporators than in dehydration equipment. Further, increased viscosity from concentration often is needed to prevent liquids from running off drying surfaces or to facilitate foaming or puffing.

Foods are also concentrated because the concentrated forms have become desirable components of diet in their own right. Thus, fruit juices plus sugar with concentration becomes jelly. The more common concentrated fruit and vegetable products include items as fruit and vegetable juices and nectars, jams and jellies, tomato paste, many types of fruit purées used by bakers, candy makers and other food manufacturers.

The level of water in virtually all concentrated foods is in itself more than enough to permit microbial growth. Yet while many concentrated foods such as non-acid fruit and vegetable purées may quickly undergo microbial spoilage unless additionally processed, such items as sugar syrups, jellies and jams are relatively "immune" to spoilage; the difference of course is in what is dissolved in the remaining water and what osmotic concentration is reached.

Methods of Concentration

a. Solar concentration. As in food dehydration, one of the simplest methods of evaporating water is with solar energy. A typical example of this method is production at farm level in developing countries of fruit pastes/ leathers (such as apricot or plum pastes).

b. Open Kettles. Some foods can be satisfactorily concentrated in open kettles that are heated by steam. This is the case for jellies and jams, tomato juices and purées and for certain types of soups. High temperatures and long concentration times should be avoided in order to reduce or eliminate damage. It is also necessary to avoid thickening and burn-on of product to the kettle wall as these gradually lower the efficiency of heat transfer and slow the concentration process. However, when the process is under control, this type of evaporation is still highly recommended for small scale operations in developing countries. It is a quite widely used system, mainly for jellies, jams and marmalades (Fig. 5.2.20). Figure 5.2.20 Open kettle

c. Flash evaporators.

d. Thin-film Evaporators.

e. Vacuum evaporators. - It is common to construct several vacuumised vessels in series so that the product moves from one vacuum chamber to the next and thereby becomes progressively more concentrated in stages. With such an arrangement the successive stages are maintained at progressively higher degrees of vacuum, and the hot water vapour produced by the first stage is used to heat the second stage, the vapour from the second stage heats the third stage and so on. In this way maximum use of heat energy is made. Such system is called a multiple effect vacuum evaporator and is illustrated in Fig. 5.2.21; it is a widely used system for concentrated tomato paste. Figure 5.2.21 Multiple effect vacuum evaporator

f. Freeze Concentration. This process has been known for many years and has been applied commercially to orange juice. However, high processing costs due largely to losses of juice occlude [unclear] to the ice crystals, have limited the number of installations to date.

g. Ultrafiltration and reverse Osmosis.

Changes from concentration

Obviously concentration that exposes food to 100° C or higher temperatures for prolonged periods can cause major changes in organoleptic and nutritional properties. Cooked foods and darkening of colour are two of the more common heat induced results which must be kept under control during a well designed process with an efficient evaporator which is still "safe" .

Microbial destruction is another type of change that may occur during concentration and will be largely dependent upon temperature. Concentration at a temperature of 100 °C or slightly above will kill many micro-organisms but cannot be depended on to destroy bacterial spores. When the food contains acid, such as fruit juices, the kill will be greater but again sterility is unlikely.

On the other hand, when concentration is done under vacuum many bacterial types not only survive the low temperatures but multiply in the concentrating equipment. It is therefore necessary to stop frequently and sanitise low temperature evaporators and where sterile concentrated foods are required, to resort to an additional preservation treatment.

Chemical Preservation

Many chemicals will kill micro-organisms or stop their growth but most of these are not permitted in foods; chemicals that are permitted as food preservatives are listed in Table 5.3.1. Chemical food preservatives are those substances which are added in very low quantities (up to 0.2%) and which do not alter the organoleptic and physico-chemical properties of the foods at or only very little.

Preservation of food products containing chemical food preservatives is usually based on the combined or synergistic activity of several additives, intrinsic product parameters (e.g. composition, acidity, water activity) and extrinsic factors (e.g. processing temperature, storage atmosphere and temperature).

This approach minimises undesirable changes in product properties and reduces concentration of additives and extent of processing treatments.

The concept of combinations of preservatives and treatments to preserve foods is frequently called the hurdle or barrier concept. Combinations of additives and preservatives systems provide unlimited preservation alternatives for applications in food products to meet consumer demands for healthy and safe foods.

Chemical food preservatives are applied to foods as direct additives during processing, or develop by themselves during processes such as fermentation. Certain preservatives have been used either accidentally or intentionally for centuries, and include sodium chloride (common salt), sugar, acids, alcohols and components of smoke. In addition to preservation, these compounds contribute to the quality and identity of the products, and are applied through processing procedures such as salting, curing, fermentation and smoking.

Table 1.3 Chemical Food Preservatives

Agent	*Acceptable Daily intake (mg/Kg body weight)*	*Commonly used levels (%)*
Lactic acid	No limit	No limit
Citric acid	No limit	No limit
Acetic acid	No limit	No limit
Sodium Diacetate	15	0.3-0.5
Sodium benzoate	5	0.03-0.2
Sodium propionate	10	0.1-0.3
Potassium sorbate	25	0.05-0.2
Methyl paraben	10	0.05-0.1
Sodium nitrite	0.2	0.01-0.02
Sulphur dioxide	0.7	0.005-0.2

Heat Preservation/Heat Processing

There are various degrees of preservation by heating; a few terms have to be identified and understood.

a. Sterilisation. By sterilisation we mean complete destruction of micro-organisms. Because of the resistance of certain bacterial spores to heat, this frequently means a treatment of at least 121° C (250° F) of wet heat for 15 minutes or its equivalent. It also means that every particle of the food must receive this heat treatment. If a can of food is to be sterilised, then immersing it into a 121° C pressure cooker or retort for the 15 minutes will not be sufficient because of relatively slow rate of heat transfer through the food in the can to the most distant point.

b. "Commercially sterile". Term describes the condition that exists in most of canned or bottled products manufactured under Good Manufacturing Practices procedures and methods; these products generally have a shelf-life of two years or more.

c. Pasteurized means a comparatively low order of heat treatment, generally at a temperature below the boiling point of water. The more general objective of pasteurization is to extend product shelf-life from a microbial and enzymatic point of view; this is the objective when fruit or vegetable juices and certain other foods are pasteurized. Pasteurization is frequently combined with another means of preservation - concentration, chemical, acidification, etc.

Blanching is a type of pasteurization usually applied to vegetables mainly to inactivate natural food enzymes. Depending on its severity, blanching will also destroy some microorganisms.

Sequence of operations employed in heat preservation of foods (fruit and vegetables, etc.)

In a simplified manner, the main operations employed in heat preservation can be described as follows:

Food preparation:	Preparation procedures will vary with the type of food. For fruit, washing, sorting, grading, peeling, cutting to size, pre-cooking and pulping operations may be employed.
Can/receptacle	This may be carried out manually or by using sophisticated filling machinery. The ratio of liquid to solid in the can must be carefully controlled and the can must not be overfilled. A headspace of 6-9 mm depth (6-8% of the container volume) above the level of food in the can is usual.
Vacuum production	This can be achieved by filling the heated product into the can, by heating the can and contents after filling, by evacuating the headspace gas in a vacuum chamber, or by injecting superheated steam into the headspace. In each case the can end is seamed on immediately afterwards.
Thermal processing	The filled sealed can must be heated to a high temperature for a sufficient length of time to ensure the destruction of spoilage micro-organisms. This is usually carried out in an autoclave or retort, in an environment of steam under pressure.
Cooling	The processed cans must be cooled in chlorinated water to a temperature of 37°C. At this temperature the heat remaining is sufficient to allow the water droplets on the can to evaporate before labelling and packing.
Labelling and packing	Labels are applied to the can body, and the cans are then packed into cases.

In principle, all these operations can also be carried out at the farm/community level using the appropriate small scale equipment, preferably only glass jars (e.g. no metal cans).

Technological Principles of Pasteurization

Physical and chemical factors which influence pasteurization process are the following:

a. temperature and time;

b. acidity of the products;

c. air remaining in containers.

Pasteurization processes. In pasteurising certain acid juices for example, there are two categories of processes:

a) Low pasteurization where pasteurization time is in the order of minutes and related to the temperature used; two typical temperature/time combinations are as following:

63° C to 65° C over 30 minutes or 75° C over 8 to 10 minutes.

Pasteurization temperature and time will vary according to:

- nature of product; initial degree of contamination;
- pasteurized product storage conditions and shelf life required.

In this first category of pasteurization processes it is possible to define three phases:

- heating to a fixed temperature;
- maintaining this temperature over the established time period (= pasteurization time);
- cooling the pasteurized products: natural (slow) or forced cooling.

b) Rapid, high or flash pasteurization is characterized by a pasteurization time in the order of seconds and temperatures of about 85° to 90° C or more, depending on holding time. Typical temperature/time combinations are as follows:

88° C (190° F) for 1 minute; 100° C for 12 seconds; 121°C for 2 seconds.

While bacterial destruction is very nearly equivalent in low and in high pasteurization processes, the 121° C/2 seconds treatment give the best quality products in respect of flavour and vitamin retention. Such short holding times, however, require special equipment which is more difficult to design and generally is more expensive than the 63-65 ° C/30 minutes type of processing equipment.

In flash pasteurization the product is heated up rapidly to pasteurization temperature, maintained at this temperature for the required time, then rapidly cooled down to the temperature for filling, which will be performed in aseptic conditions in sterile receptacles. Taking into account the short time and rapid performance of this operation, flash pasteurization can only be achieved in continuous process, using heat exchangers. Industrial applications of pasteurization process are mainly used as a means of preservation for fruits and vegetable juices and specially for tomato juice. Thermopenetration. The thermopenetration problem is extremely important, especially in the case of the pasteurization of products packed in glass containers because it is the determining factor for the success of the whole operation.

During pasteurization it is necessary that a sufficient heat quantity is transferred through the receptacle walls; this is in order that the product temperature rises sufficiently to be lethal to micro-organisms throughout the product mass.

The most suitable and practical method to speed up thermopenetration is the movement of receptacles during the pasteurization process. Rapid rotation of receptacles around their axis is an efficient means to accelerate heat transfer, because this has the effect, among others of rapidly mixing the contents.

The critical speed of for this movement is generally about 70 rotations per minute (RPM). This enables a more uniform heating of products, reducing heating time and organoleptic degradation.

Heating may precede or follow packaging. These principles of different temperature time combinations very largely determine the design parameters for heat preservation equipment and commercial practices.

The food processor will employ no less than that heat treatment which gives the necessary degree of micro-organism destruction. This is further ensured by periodic inspection by local sanitary authorities or by the importing countries sanitary services. However, the food processor also will want to use the mildest effective heat treatment to ensure highest food quality.

It is convenient to separate heat preservation practices into two broad categories: one involves heating of foods in their final containers, the other employs heat prior to packaging.

The latter category includes methods that are inherently less damaging to food quality, where the food can be readily subdivided (such as liquids) for rapid heat exchange. However, these methods then require packaging under aseptic or nearly aseptic conditions to prevent or at least minimise recontamination.

On the other hand, heating within the package frequently is less costly and produces quite acceptable quality with the majority of foods and most of our present canned food supply is heated in the package.

In practice, therefore, most of the canned food produced locally in developing countries should be heated within the package.

Simplified and illustrated flow-sheet of the operation cycle for fruit and vegetables heat preservation) shows a simplified and illustrated flow-sheet of the operation cycle for fruit and vegetable heat preservation.

Food Irradiation

Food irradiation is one of the food processing technologies available to the food industry to control organisms that cause food-borne diseases and to reduce food losses due to spoilage and deterioration. Food irradiation technology offers some advantages over conventional processes. Each application should be evaluated on its own merit as to whether irradiation provides a technical and economical solution that is better than traditional processing methods.

Applications

For products where irradiation is permitted, commercial applications depend on a number of factors including the demand for the benefits provided, competitiveness with alternative processes and the willingness of consumers to buy irradiated food products. There are a number of applications of food irradiation. For each application it is important to determine the optimum dosage range required to achieve the desired effect. Too high a dosage can produce undesirable changes in texture, colour and taste of foods.

Shelf-life extension. Irradiation can extend the shelf-life of foods in a number of ways. By reducing the number of spoilage organisms (bacteria, mould, fungi), irradiation can lengthen the shelf life of fruits and vegetables.

Since ionising radiation interferes with cell division, it can be used as an alternative to chemicals to inhibit sprouting and thereby extend the shelf life of potatoes, onions and garlic. Exposure of fruits and vegetables to ionising radiation slows their rate of ripening. Strawberries, for example, have been found to be suitable for irradiation. Their shelf-life can be extended three-fold, from 5 to 15 days.

Fruit Specific Preservation Technologies

Fruit quality

Fruit quality goes back to tree stock, growing practices and weather conditions. Closer to the shipper and processor, however, are the degrees of maturity and ripeness when picked and the method of picking or harvesting. There is a distinction between maturity and ripeness of a fruit. Maturity is the condition when the fruit is ready to eat or if picked will become ready to eat after further ripening. Ripeness is that optimum condition when colour, flavour and texture have developed to their peak.

Some fruit is picked when it are mature but not yet ripe. This is especially true of very soft fruit like cherries and peaches, which when fully ripe are so soft as to be damaged by the act of picking itself. Further, since many types of fruit continue to ripen off the tree, unless they were to be processed quickly, some would become overripe before they could be utilised if picked at peak ripeness.

From a technological point of view, fruit characterisation by species and varieties is performed on the basis of physical as well chemical properties: shape, size, texture, flavour, colour/pigmentation, dry matter content (soluble solids content), pectic substances, acidity, vitamins, etc. These properties are directly correlated with fruit utilisation. Table 8.1. 1 shows which of the above mentioned properties have a major impact on the finished products obtained by fruit processing.

When to pick

The proper time to pick fruit depends upon several factors; these include variety, location, weather, ease of removal from the tree (which change with time), and purpose to which the fruit will be put.

For example, oranges change with respect to both sugar and acid as they ripen on the tree; sugar increases and acid decreases. The ratio of sugar to acid determines the taste and acceptability of the fruit and the juice. For this reasons, in some countries there are laws that prohibit picking until a certain sugar-acid ratio has been reached.

In the case of much fruit to be canned, on the other hand, fruit is picked before it is fully ripe for eating since canning will further soften the fruit.

Quality measurements

Many quality measurements can be made before a fruit crop is picked in order to determine if proper maturity or degree of ripeness has developed. Colour may be measured with instruments or by comparing the colour of fruit on the tree with standard picture charts. Texture may be measured by compression by hand or by simple type of plungers.

As fruit mature on the tree its concentration of juice solids, which are mostly sugars, changes. The concentration of soluble solids in the juice can be estimated with a refractometer or a hydrometer. The refractometer measures the ability of a solution to bend or refract a light beam which is proportional to the solution's concentration. A hydrometer is a weighted spindle with a graduated neck which floats in the juice at a height related to the juice density.

The acid content of fruit changes with maturity and affects flavour. Acid concentration can be measured by a simple chemical titration on the fruit juice. But for many fruits the tartness and flavour are really affected by the ratio of sugar to acid.

Percentage of soluble solids, which are largely sugars, is generally expressed in degrees Brix, which relates specific gravity of a solution to an equivalent concentration of pure sucrose.

In describing the taste of tartness of several fruits and fruit juices, the term "sugar to acid ratio" or "Brix to acid ratio" are commonly used. The higher the Brix the greater the sugar concentration in the juice; the higher the "Brix to acid ratio" the sweeter and lees tart is the juice.

Harvesting

The above and other measurements, plus experience, indicate when fruit is ready for harvesting and subsequent processing.

A large amount of the harvesting of most fruit crops is still done by hand; this labour may represent about half of the cost of growing the fruit. Therefore, mechanical harvesting is currently one of the most active fields of research for the agricultural engineer, but also requires geneticists to breed fruit of nearly equal size, that matures uniformly and that is resistant to mechanical damage.

A correct manual harvesting includes some simple but essential rules:

- the fruit should be picked by hand and placed carefully in the harvesting basket; all future handling has to be performed carefully in order to avoid any mechanical damage;
- the harvesting basket and the hands of the harvester should be clean;
- the fruit should be picked when it is ready to be able to be processed into a quality product depending on the treatment which it will undergo.

It is worth emphasising the fact that the proximity of the processing centre to the source of supply for fresh raw materials presents major advantages; some are as follows:

- possibility to pick at the best suitable moment;
- reduction of losses by handling/transportation;
- minimises raw material transport costs;
- possibility to use simpler/cheaper receptacles for raw material transport.

Once it has left the tree, the organoleptic properties, nutritional value, safety and aesthetic appeal of the fruit deteriorates in varying degrees. The major causes of deterioration include the following:

a. growth and activity of micro-organisms;

b. activities of the natural food enzymes;

c. insects, parasites and rodents;

d. temperature, both heat and cold;

e. moisture and dryness;

f. air and in particular oxygen;

g. light and

h. time.

Reception - quality and quantity

Fruit reception at the processing centre is performed mainly for following purposes:

- checking of sanitary and freshness status;
- control of varieties and fruit wholeness;
- evaluation maturity degree;
- collection of data about quantities received in connection to the source of supply: outside growers/farmers, own farm.

Variety control is needed in order to identify that the fruit belongs to an accepted variety as not all are suitable for different technological processes.

Fruit maturity degree is significant as industrial maturity is required for some processing/preservation methods while for others there is the need for an edible maturity when the fruit has full taste and flavour.

Special attention is given to size, appearance and uniformity of fruit to be processed, mainly in the form of fruit preserved with sugar using whole/half fruits ("with fruit pieces").

Some laboratory control is also needed, even if it not easy to precisely establish the technological qualities of fruit because of the absence of enough reliable rapid analytical methods able to show eventual deterioration.

The only reliable method for evaluating the quality is the combination of data obtained through organoleptic/taste controls and by simple analytical checks which are possible to perform in a small laboratory: percentage of soluble solids by refractometer, consistency/texture measured with simple penetrometers, etc.

Temporary storage before processing

This step has to be as short as possible in order to avoid flavour losses, texture modification, weight losses and other deterioration that can take place over this period.

Some basic rules for this step are as follows:

- keep products in the shade, without any possible direct contact with sunlight;
- avoid dust as much as possible;
- avoid excessive heat;
- avoid any possible contamination;
- store in a place protected from possible attack by rodents, insects, etc.

Cold storage is always highly preferred to ambient temperature. For this reason a very good manufacturing practice is to use a cool room for each processing centre; this is very useful for small and medium processing units as well.

Washing

Harvested fruit is washed to remove soil, micro-organisms and pesticide residues.

Fruit washing is a mandatory processing step; it would be wise to eliminate spoiled fruit before washing in order to avoid the pollution of washing tools and/ or equipment and the contamination of fruit during washing.

Washing efficiency can me gauged by the total number of micro-organisms present on fruit surface before and after washing - best result are when there is a six fold reduction. The water from the final wash should be free from moulds and yeast; a small quantity of bacteria is acceptable.

Fruit washing can be carried out by immersion, by spray/ showers or by combination of these two processes which is generally the best solution: pre-washing and washing.

Some usual practices in fruit washing are:

- addition of detergents or 1.5% HCl solution in washing water to remove traces of insect-fungicides;
- use of warm water (about 50°C) in the pre-washing phase;
- higher water pressure in spray/shower washers.

Washing must be done before the fruit is cut in order to avoid losing high nutritive value soluble substances (vitamins, minerals, sugars, etc.).

Sorting

Fruit sorting covers two main separate processing operations:

a. removal of damaged fruit and any foreign bodies (which might have been left behind after washing);

b. qualitative sorting based on organoleptic criteria and maturity stage.

Mechanical sorting for size is usually not done at the preliminary stage. The most important initial sorting is for variety and maturity.

However, for some fruit and in special processing technologies it is advisable to proceed to a manual dimensional sorting (grading).

Trimming and peeling (skin removal)

This processing step aims at removing the parts of the fruit which are either not edible or difficult to digest especially the skin.

Up to now the industrial peeling of fruit and vegetables was performed by three procedures:

a. mechanically;

b. by using water steam;

c. chemically; this method consists in treating fruit and vegetables by dipping them in a caustic soda solution at a temperature of 90 to 100° C; the concentration of this solution as well as the dipping or immersion time varying according to each specific case.

Cutting

This step is performed according to the specific requirements of the fruit processing technology.

Heat blanching

Fruit is not usually heat blanched because of the damage from the heat and the associated sogginess and juice loss after thawing. Instead, chemicals are commonly used without heat to inactivate the oxidative enzymes or to act as antioxidants and they are combined with other treatments.

Ascorbic/citric acid dip

Ascorbic acid or vitamin C minimises fruit oxidation primarily by acting as an antioxidant and itself becoming oxidised in preference to catechol-tannin compounds. Ascorbic acid is frequently used by being dissolved in water, sugar syrup or in citric acid solutions.

It has been found that increased acidity also helps retard oxidative colour changes and so ascorbic acid plus citric acid may be used together. Citric acid further reacts with (chelates) metal ions thus removing these catalysts of oxidation from the system.

Sulphur dioxide treatment

Sulphur dioxide may function in several ways:

- sulphur dioxide is an enzyme poison against common oxidising enzymes;
- it also has antioxidant properties; i.e., it is an oxygen acceptor (as is ascorbic acid);
- further SO_2 minimises non enzymatic Maillard type browning by reacting with aldehyde groups of sugars so that they are no longer free to combine with amino acids;
- sulphur dioxide also interferes with microbial growth.

In many fruit processing pre-treatments two factors must be considered:

a. sulphur dioxide must be given time to penetrate the fruit tissues;

b. SO_2 must not be used in excess because it has a characteristic unpleasant taste and odour, and international food laws limit the SO_2 content of fruit products, especially of those which are consumer oriented (*e.g.* except semi-processed products oriented to further industrial utilisation).

Commonly a 0.25 % solution (except for semi-processed fruit products which are industry oriented and use a 6% solution) of SO_2 or its SO_2 equivalent in the form of solutions of sodium sulphite, sodium bisulphite or sodium/potassium metabisulphite are used.

Fruit slices are dipped in the solution for about two to three minutes and then removed so as not to absorb too much SO_2. Then the slices are allowed to stand for about one to two hours so that the SO_2 may penetrate throughout the tissues before processing.

Sulphur dioxide is also used in fruit juice production to minimise oxidative changes where relatively low heat treatment is employed so as not to damage delicate juice flavour.

Dry sulphuring is the technological step where fruit is exposed to fumes of SO_2 from burning sulphur or from compressed gas cylinders; this treatment could be used in the preparation of fruits (and some vegetables) prior to drying / dehydration.

Sugar syrup

Sugar syrup addition is one of the oldest methods of minimising oxidation. It was used long before the causative reactions were understood and remains today a common practice for this purpose.

Sugar syrup minimises oxidation by coating the fruit and thereby preventing contact with atmospheric oxygen.

Sugar syrup also offers some protection against loss of volatile fruit esters and it contributes sweet taste to otherwise tart fruits.

It is common today to dissolve ascorbic acid and citric acid in the sugar syrup for added effect or to include sugar syrup after an SO_2 treatment.

Fresh fruit storage

Some fruit species and specially apples and pears can be stored in fresh state during cold season in some countries' climatic conditions.

Fruit for fresh storage have to be autumn or winter varieties and be harvested before they are fully mature. This fruit also has to be sound and without any bruising; control and sorting by quality are mandatory operations.

Sorting has to be carried out according to size and weight and also by appearance; fruit which is not up to standard for storage will be used for semi-processed product manufacturing which will be submitted further to industrial processing.

Harvested fruit has to be transported as soon as possible to storage areas. Leaving fruit in bulk in order to generate transpiration is a bad practice as this reduces storage time and accelerates maturation processes during storage.

Vegetable Specific Processing Technologies

Vegetables Varieties

Vegetable processors must appreciate the substantial differences that varieties of a given vegetable will possess. In addition to variety and genetic strain differences with respect to weather, insect and disease resistance, varieties of a given vegetable will differ in size, shape, time of maturity, and resistance to physical damage.

Varietal differences then further extend into warehouse storage stability, and suitability for such processing methods as canning, freezing, pickling or drying. A variety of peas that is suitable for canning may be quite unsatisfactory for freezing and varieties of potatoes that are preferred for freezing may be less satisfactory for drying or potato chip manufacture.

This should be expected since different varieties of a given vegetable will vary somewhat in chemical composition, cellular structure and biological activity of their enzyme system.

Harvesting and Pre-processing

When vegetables are maturing in the field they are changing from day to day. There is a time when the vegetable will be at peak quality from the stand-point of colour, texture and flavour.

This peak quality is quick in passing and may last only a day. Harvesting and processing of several vegetables, including tomatoes, corn and peas are rigidly scheduled to capture this peak quality.

After the vegetable is harvested it may quickly pass beyond the peak quality condition. This is independent of microbiological spoilage; these main deteriorations are related to:

a) loss of sugars due to their consumption during respiration or their conversion to starch; losses are slower under refrigeration but there is still a great change in vegetable sweetness and freshness of flavour within 2 or 3 days;

b) production of heat when large stockpiles of vegetables are transported or held prior to processing.

At room temperature some vegetables will liberate heat at a rate of 127,000 kJ/ton/day; this is enough for each ton of vegetables to melt 363 kg of ice per day. Since the heat further deteriorates the vegetables and speeds micro-organisms growth, the harvested vegetables must be cooled if not processed immediately. But cooling only slows down the rate of deterioration, it does not prevent it, and vegetables differ in their resistance to cold storage. Each vegetable has its optimum cold storage temperature which may be between about 0-100°C (32-500°F).

c) the continual loss of water by harvested vegetables due to transpiration, respiration and physical drying of cut surfaces results in wilting of leafy vegetables, loss of plumpness of fleshy vegetables and loss of weight of both.

Moisture loss cannot be completely and effectively prevented by hermetic packaging. This was tried with plastic bags for fresh vegetables in supermarkets but the bags became moisture fogged, and deterioration of certain vegetables was accelerated because of buildup of CO_2 and decrease of oxygen in the package. It therefore is common to perforate such bags to prevent these defects as well as to minimise high humidity in the package which would encourage microbial growth.

Shippers of fresh vegetables and vegetable processors, whether they can, freeze, dehydrate, or manufacture soups or ketchup, appreciate the instability and perishability of vegetables and so do everything they can to minimise delays in processing of the fresh product. In many processing plants it is common practice to process vegetables immediately as they are received from the field.

To ensure a steady supply of top quality produce during the harvesting period the large food processors will employ trained field men; they will advise on growing practices and on spacing of plantings so that vegetables will mature and can be harvested in rhythm with the processing plant capabilities. This minimises stockpiling and need for storage.

Cooling of vegetables in the field is common practice in some areas. Liquid nitrogen-cooled trucks may next provide transportation of fresh produce to the processing plant or directly to market.

Upon arrival of vegetables at the processing centre the usual operations of cleaning, grading, peeling, cutting and the like are performed using a moderate amount of equipment but a good deal of hand labour also still remains.

Reception

This covers qualitative and quantitative control of delivered vegetables. The organoleptic control and the evaluation of the sanitary state, even if they are very important steps in vegetables' characteristics assessment, cannot establish their technological value.

On the other hand, laboratory controls do not precisely establish their technological properties because of the difficulty in putting into showing some deterioration when using rapid control methods.

One correct method of vegetable quality appraisal is their overall evaluation based on the whole complex of data that can be obtained by combining an extensive organoleptic evaluation with simple analysis that can be performed rapidly in plant laboratory. These analysis can be:

a. refractometric extract (tomatoes, fruit, etc.);

b. specific weight (potatoes, peas, etc.);

c. consistency (measured with tenderometers, penetrometers, etc.);

d. boiling tests, etc.

Temporary Storage

This step should be as short as possible and better completely eliminated. Vegetables can be stored in:

a. simple stores, without artificial cooling;

b. in refrigerated stores; or, in some cases,

c. in silos (potatoes, etc.).

Simple stores should be covered, fairly cool, dry and well ventilated but without forced air circulation which can induce significant losses in weight through intensive water evaporation; air relative humidity should be at about 70-80%.

Refrigerated storage is always preferable and in all cases a processing centre needs a cold room for this purpose, adapted in volume I capacity to the types and quantities of vegetables (and fruits) that are further processed.

Washing

Washing is used not only to remove field soil and surface micro-organisms but also to remove fungicides, insecticides and other pesticides, since there are laws specifying maximum levels of these materials that may be retained on the vegetable; and in most cases the allowable residual level is virtually zero. Washing water contains detergents or other sanitisers that can essentially completely remove these residues.

The washing equipment, like all equipment subsequently used, will depend upon the size, shape and fragility of the particular kind of vegetable:

- flotation cleaner for peas and other small vegetables;
- rotary washer in which vegetables are tumbled while they are sprayed with jets of water; this type of washer should not be used to clean fragile vegetables;

Sorting

This step covers two separate operations:

a) removal of non-standard vegetables (and fruit) and possible foreign bodies remaining after washing;

b) quality grading based on variety, dimensional, organoleptical and maturity stage criterion.

Skin Removal/peeling

Some vegetables require skin removal. This can be done in various ways.

a) Mechanical

This type of operation is performed with various types of equipment which depend upon the result expected and the characteristics of the fruit and vegetables, for example:

i. a machine with abrasion device (potatoes, root vegetables);

ii. equipment with knives (apples, pears, potatoes, etc.);

iii. equipment with rotating sieve drums (root vegetables). Sometimes this operation is simultaneous with washing (potatoes, carrots, etc.) or preceded by blanching (carrots).

b) Chemical

Skins can be softened from the underlying tissues by submerging vegetables in hot alkali solution. Lye may be used at a concentration of about 0.5-3%, at about 93°C (2000°F) for a short time period (0.5-3 min). The vegetables with loosened skins are then conveyed under high velocity jets of water which wash away the skins and residual lye.

In order to avoid enzymatic browning, this chemical peeling is followed by a short boiling in water or an immersion in diluted citric acid solutions.

It is more difficult to peel potatoes with this method because it is necessary to dissolve the cutin and this requires more concentrated lye solutions, up to 10%.

c) Thermal

Wet heat (steam). Other vegetables with thick skins such as beets, potatoes, carrots and sweet potatoes may be peeled with steam under pressure (about 10 at) as they pass through cylindrical vessels. This softens the skin and the underlying tissue. When the pressure is suddenly released, steam under the skin expands and causes the skin to puff and crack. The skins are then washed away with jets of water at high pressure (up to 12 at).

Dry heat (flame). Other vegetables such as onions and peppers are best skinned by exposing them to direct flame (about 1 min at 1000°C) or to hot gases in rotary tube flame peelers. Here too, heat causes steam to develop under skins and puff them so that they can be washed away with water.

Manual peeling only use when the other methods are impossible or sometimes as a completion of the other three ways. Average losses at this step are given in Table 9.2.1.

Blanching

The special heat treatment to inactivate enzymes is known as blanching. Blanching is not indiscriminate heating. Too little is ineffective, and too much damages the vegetables by excessive cooking, especially where the fresh character of the vegetable is subsequently to be preserved by processing.

This heat treatment is applied according to and depends upon the specificity of vegetables, the objectives that are followed and the subsequent processing / preservation methods.

Two of the more heat resistant enzymes important in vegetables are catalase and peroxidase. If these are destroyed then the other significant enzymes in vegetables also will have been inactivated. The heat treatment to destroy catalase and peroxidase in different vegetables are known, and sensitive chemical tests have been developed to detect the amounts of these enzymes that might survive a blanching treatment. Catalase and peroxidase inactivation tests are presented in section 9.2.9.

Because various types of vegetables differ in size, shape, heat conductivity, and the natural levels of their enzymes, blanching treatments have to be established on an experimental basis. As with sterilisation of foods in cans, the larger the food item the longer it takes for heat to reach the centre. Small vegetables may be adequately blanched in boiling water in a minute or two, large vegetables may require several minutes.

Blanching as a unit operation is a short time heating in water at temperatures of 100°C or below. Water blanching may be performed in double bottom kettles, in special baths with conveyor belts or in modern continuous blanching equipment.

In order to reduce losses of hydrosoluble substances (mineral salts, vitamins, sugars, etc.) occurring during water blanching, several methods have been developed:

- temperature setting at 85-95°C instead of 100°C;
- blanching time has to be just sufficient to inactivate enzymes catalase and peroxidase;
- assure elimination of air from tissues.

Table 1.4 Blanching parameters for some vegetables

Vegetables	*Temperature, °C*	*Time, min.*
Peas	85-90	2-7
Green beans	90-95	2-5
Cauliflower	Boiling	2
Carrots	90	3-5
Peppers	90	3

Steam heat treatment can also be applied instead of water blanching as a preliminary step before freezing or drying, as long as the preservation method is only used for enzyme inactivation and not to modify consistency.

For drying, the vegetables are conveyed directly from steaming equipment to drying installations without cooling. Vegetable steaming is carried out in continuous installations with conveyer belts made from metallic sieves.

Cooling of vegetables after water blanching or steaming is performed in order to avoid excessive softening of the tissues and has to follow immediately after these operations; one exception is the case of vegetables for drying which can be transferred directly to drying equipment without cooling.

Natural cooling is not recommended because is too long and generates significant losses in vitamin C content. Cooling in pre-cooled air (from special installations) is sometimes used for vegetables that will be frozen

Cooling in water can be achieved by sprays or by immersion; in any case the vegetables have to reach a temperature value under 37°C as soon as possible. Too long a cooling time generates supplementary losses in valuable hydrosoluble substances; in order to avoid this, the temperature of the cooling water has to be as low as possible.

Canning

Large quantities of vegetable products are canned. A typical flow sheet for a vegetable canning operation (which also applies to fruit for the most part) covers some food process unit operations performed in sequence: harvesting; receiving; washing; grading; heat blanching; peeling and coring; can filling; removal of air under vacuum; sealing/closing, retorting/heat treatment; cooling; labelling and packing. The vegetable may be canned whole, diced, puréed, as juice and so on.

Fresh Vegetable Storage

The vegetables can be stored, in some specific natural conditions, in fresh state, that is without significant modifications of their initial organoleptic properties. Fresh vegetable storage can be short term; this was briefly covered under temporary storage before processing. Also fresh vegetable storage can be long term during the cold season in some countries and in this case it is an important method for vegetable preservation in the natural state.

In order to assure preservation in long term storage, it is necessary to reduce respiration and transpiration intensity to a minimum possible; this can be achieved by:

a. maintenance of as low a temperature as possible (down to 0°C),

b. air relative humidity increased up to 85-95 % and

c. CO_2 percentage in air related to the vegetable species.

Vegetables for storage must conform to following conditions: they must be of one of the autumn or winter type variety; be at edible maturity without going past this stage; be harvested during dry days; be protected from rain, sun heat or wind; be in a sound state and clean from soil; be undamaged.

From the time of harvest and during all the period of their storage vegetables are subject to respiration and transpiration and this is on account of their reserve substances and water content. The more the intensity of these two natural processes are reduced, the longer sound storage time will be and the more losses will be reduced.

For this reason, vegetables have to be handled and transported as soon as possible in the storage conditions (optimal temperature and air relative humidity for the given species). Even in these optimal conditions storage will generate losses in weight which are variable and depend upon the species.

Some optimal storage conditions are shown in table below.

Table 1.5 Optimal conditions for fresh vegetable storage

Vegetables	*Storage conditions*	
	Temperature, °C	*Relative humidity, %*
Potatoes	+1...+3	85-90
Carrots	0 ... +1	90-95
Onions	0 ... +1	75-85
Leeks	0 ... +0.5	85-90
Cabbage	-1 ... 0	90-97
Garlic	0 ... +1	85-90
Beets	0 ... +1	90-95

Technology for Vegetable Powder Processing

This technology has been developed in recent years with applications mainly for potatoes (flour, flakes, granulated), carrots (powder) and red tomatoes (powder). In order to obtain these finished products there are two processes:

a) drying of vegetables down to a final water content below 4% followed by grinding, sieving and packing of products;

b) vegetables are transformed by boiling and sieving into purées which are then dried on heated surfaces (under vacuum preferably) or by spraying in hot air.

Industrial installations that can be used for these products and technological data are summarised below:

- Dryers with plates under vacuum are equipped with plates heated with hot water. Stainless steel plates containing the purée to be dried are placed on them. Process conditions are at low residual pressure (about 10-20 mm Hg) and a product temperature of 50-70°C. This equipment is discontinuous but easy to operate.
- Drum dryers have one or two drums heated with hot water or steam as heating elements. Feeding is continuous between the two drums which are rotating in reverse direction (about 2-6 rotations per minute) and the distance of which is adjustable and determines the thickness of layer to be dried. he product is dried and removed by mechanical means during rotation.
- Drying installations by spraying in hot air; the product is introduced in equipment and sprayed by a special device in hot air. Drying is instantaneous (1/50 s) and therefore can be carried out at 130-15O°C.

Packing and Storage of Dried and Powdered Vegetables

Dried vegetables can suffer significant modifications that bring about their deterioration during storage. The factors that determine these degradations impose at same time the type of packaging materials and storage conditions for packaged products. The main factor in maintaining the quality of dried products is to follow the maximum moisture contents that have to be as close as possible to the limits indicated in Table 9.4.4. The moisture content of dried vegetables is not constant because of their hygroscopicity and is always in equilibrium with relative humidity of air in storage rooms. Technical solutions for maintaining a low dehydrated products moisture are:

a) storage in stores with air relative humidity below 78%;

b) use packages that are water vapour proof. The most efficient packages are tin boxes or drums (mainly for long term storage periods); combined packages (boxes, bags, etc.) from complexes (carton with metallic sheets, plastic materials, etc.) mainly for small packages. One solution for some dried vegetables may be the use of waterproof plywood drums. Modern solutions are oriented not only to the maintaining product moisture during storage but also reducing this parameter by the use of desiccants (substances which absorb moisture) introduced in packages, hermetically closed.

A desiccant in current use is calcium oxide. Granulated calcium oxide is introduced in small bags from a material which is permeable to water vapour but which does not permit the desiccant to escape into products. With desiccants, product moisture can be reduced to even below 4%, and this inhibits or reduces the biochemical and microbiological processes during storage.

Another factor that can deteriorate dried/dehydrated vegetables is atmospheric oxygen through the oxidative phenomena that it produces. In order to eliminate the action of this agent some packing methods under vacuum or in inert gases (carbon dioxide or nitrogen) are in use, applied mainly for packing dried carrots in order to avoid beta-carotene oxidation in beta-ionone (foreign smell, discoloration, etc.). In order to avoid the action of oxygen it is also possible to add ascorbic acid as antioxidant (for example in carrot powder). Sun or artificial light action on dehydrated vegetables generally causes discoloration which can be avoided by using opaque packaging materials.

Dehydrated vegetable compression (especially for roots) to form blocks with a weight of 50-600 g, is practiced sometimes; it has as advantages the reduction of evaporation surface and contact with atmospheric oxygen and volume reduction. Dehydrated vegetables are compressed at about 300 at. Compressed blocks are packaged in heat sealed plastic materials.

Storage temperature has an important role because this reduces or inhibits the speed of all physico-chemical, biochemical and microbiological processes, and thus prolongs storage period. The storage temperature should be below 25°C (and preferably 15°C); lower temperatures (0-10°C) help maintain taste, colour and water rehydration ratio and also, to some extent, vitamin C.

Chapter 2

Milk and Milk Products Processing

Although cow's milk is the most popular in many countries, milk can be obtained from many different sources. For example, milk from goats and sheep makes a substantial contribution to the total milk production in countries of Eastern and Southern Europe, Malawi, and Barbados, whereas the water buffalo is a common source of milk in much of Asia. The table below illustrates some of the differences in composition between these milks.

Milk is a perishable commodity and spoils very easily. Its low acidity and high nutrient content make it the perfect breeding ground for bacteria, including those which cause food poisoning (pathogens).

Bacteria from the animal, utensils, hands, and insects may contaminate the milk, and their destruction is the main reason for processing. This preservation of the milk can be achieved by fermentation, heating, cooling, removal of water, and by concentration or separation of components, to produce foods such as butter or cheese.

The degree to which milk consumption and processing occurs will differ from region to region. It is dependent upon a whole host of factors, including geographic and climatic conditions, availability and cost of milk, food taboos, and religious restrictions. Where processing does exist, many traditional techniques can be found for producing indigenous milk products. These are more stable than raw milk and provide a means of preservation as well as adding variety to the diet. In addition, the introduction of western-style dairy products and the subsequent setting up of small-scale dairies has provided more choice of dairy products to the consumer.

Nutritional Significance

Milk is often regarded as being nature's most complete food. It earns this reputation by providing many of the nutrients which are essential for the growth of the human body. Being an excellent source of protein and having an abundance of vitamins and minerals, particularly calcium, milk can make a positive contribution to the health of a nation. The realization of its nutritional attributes is clearly illustrated by the implementation of numerous 'school milk programmes' worldwide.

Fermented-milk products such as yoghurt and soured milk contain bacteria from the *Lactobacilli* group. These bacteria occur naturally in the digestive tract and have a cleansing and healing effect. Therefore the introduction of fermented products into the diet can help prevent certain yeasts and bacteria which may cause illness.

Many people suffer from a condition known as 'lactose intolerance'. This means that they are unable to digest the milk fat (lactose). Such people can, however, tolerate milk if it is fermented to produce foods such as yoghurt. During fermentation, lactic acid producing bacteria break down lactose, and in doing so eliminate the cause of irritation.

The Quality of Milk

The type of animal, its quality, and its diet can lead to differences in the colour, flavour, and composition of milk. Infections in the animal which cause illness may be passed directly to the consumer through milk. It is therefore extremely important that quality-control tests are carried out to ensure that the bacterial activity in raw milk is of an acceptable level, and that no harmful bacteria remain in the processed products.

Table 2.1 Average composition (%) of milks of various mammals

Species	*Water*	*Fat*	*Protein*	*Lactose*	*Ash*
Human	87.43	3.75	1.63	6.98	0.21
Cow	87.2	3.7	3.5	4.9	0.7
Goat	87.00	4.25	3.52	4.27	0.86
Sheep	80.71	7.9	5.23	4.81	0.9
Indian buffalo	82.76	7.38	3.6	5.48	0.78
Camel	87.61	5.38	2.98	3.26	0.7
Horse	89.04	1.59	2.69	6.14	0.51
Llama	86.55	3.15	3.9	5.6	0.8

Standard Testing Procedures

Milk Fat

The price paid for milk is usually dependent upon the milk-fat content, and this may be determined either at the collection stage or at the dairy using a piece of equipment known as a butyrometer. Additionally the specific gravity can be measured using a hydrometer. This can also be used as an aid to detect adulteration.

Bacterial activity

Routinely it is necessary to check the microbiological quality of raw milk using either methylene blue or resazurin dyes. These tests indicate the activity of bacteria in the milk sample and the results determine whether the milk is accepted or rejected.

Both tests work on the principle of the time taken to change the colour of the dye. The length of time taken is proportional to the number of micro-organisms present (the shorter the time taken, the higher the bacterial activity). It is preferable to use the resazurin test as this is less time-consuming. For these tests, basic laboratory equipment will be needed such as test-tubes, a water bath, accurate measuring equipment, and a supply of dyes. After collection the milk should ideally be stored at a temperature of 4°C or below. This is necessary to slow the growth of any contaminating bacteria.

Phosphatase test

For pasteurized milk, it is possible to ensure that pasteurization has been adequately achieved by testing for the presence of the enzyme phosphatase. The destruction of phosphatase is regarded as a reliable test to show that the milk has been sufficiently heat-processed, because this enzyme (present in raw milk) is destroyed by pasteurization conditions.

It is stressed that pasteurization is an effective safeguard against spoilage and food poisoning only if the milk is not re-contaminated after pasteurization.

Processing

Liquid Milk

Milk can be kept for longer periods of time if it is heated to destroy the bacteria or cooled to slow their growth. Pasteurization and sterilization are the two most commonly-used heat treatments. Technically, it is possible for both to be carried out on a small scale, but they are most usually performed on a larger industrial scale due to the need for qualified, experienced staff and accurate and strictly controlled hygienic processing conditions.

Production stages for pasteurized and sterilized milk

Homogenization

Homogenization breaks up the oil droplets in milk and prevents the cream from separating out and forming a layer. This is of particular importance for sterilized milk which has a long shelf-life and when the formation of a cream layer is not desired. Additional changes include increased viscosity and a richer taste. Homogenizers are more usually designed for industrial-scale production, but it is possible to purchase smaller versions.

Filling

The most common packaging material for both pasteurized and sterilized milk is glass bottles sealed with either foil or metal caps, although plastic bottles, plastic bags, and cardboard cartons are all used when bottles are not available or too expensive.

Pasteurization

Pasteurization is a relatively mild heat treatment, (usually performed below 100°C) which is used to extend the shelf-life of milk for several days. It preserves the milk by the inactivation of enzymes and destruction of heat-sensitive micro-organisms, but causes minimal changes to the nutritive value or sensory characteristics of a food. Some heat-resistant bacteria survive to spoil the milk after a few days, but these bacteria do not cause food poisoning.

The time and temperature combination needed to destroy 'target' microorganisms will vary according to a number of complex inter-related factors. For milk, the heating time and temperature is either 63°C for 30 minutes or alternatively 72°C for 15 seconds. Only the former combination is possible on a small scale and for this the simplest equipment required is an open boiling pan. Better control is achieved using a steam jacketed pan, and this can be fitted with a stirrer to improve the efficiency of heating. Both of these are batch processes which are suited to small-scale operation. A higher production rate may be possible using a tubular-coil pasteurizer. This equipment has been tested and has been successful for some fruit products but it is presently still at a developmental stage.

Sterilization

Sterilization is a more severe heat treatment designed to destroy all contaminating bacteria. The milk is sterilized at a temperature of 121°C maintained for 15-20 minutes. This can be achieved using a retort or pressure cooker. Unlike pasteurization, this process causes substantial changes to the nutritional and sensory quality of the milk. In some countries, flavoured milk has become a very popular product.

However, sterilization is not recommended for small-scale production for the following reasons:

- The cost of a retort and ancillary equipment is high for the small-scale processor.
- It is essential that the correct heating conditions are carefully established and maintained for every batch of milk that is processed. If the milk is overheated, the quality is reduced, and it may have a rather burnt taste and aroma.
- If the milk is not heated sufficiently, there is a risk that micro-organisms will survive and grow inside the bottle. In low-acid foods such as milk, many types of bacteria including *Clostridium botulinum* can grow and cause severe food poisoning.
- Due to the potential dangers from food poisoning, the skills of a qualified food technologist/microbiologist are required in order to routinely

examine samples of sterilized milk that have been subjected to accelerated storage conditions. This requires a supply of microbiological media and equipment.

In summary, the process of sterilization requires a considerable capital investment, the need for trained and experienced staff, regular maintenance of sophisticated equipment, and a comparatively high operating expenditure.

Cooling

Pasteurization does not destroy all of the micro-organisms, therefore the milk has to be cooled rapidly to prevent the growth of surviving bacteria. Cooling can be achieved on a small scale by using a bottle-cooling system. A system is outlined in the Packaging chapter.

Storage

Pasteurized milk has a shelf-life of 2-3 days if kept at 4°C. Maintaining this low temperature causes a substantial increase to the cost of transportation and distribution and is therefore a major disadvantage to the development of a small-scale pasteurized milk business. If packaged in sealed bottles and stored at room temperature, sterilized milk should have a shelf-life in excess of six months.

Separation of milk components

Cream

When milk is left to stand for some time, fat globules rise to the surface forming a layer of fat (or cream). This can be separated leaving behind skimmed milk as a by-product. There are different types of cream each with different fat concentrations: single (or light) cream contains 18 per cent milk fat whereas double (or heavy) cream normally contains 30 per cent milk fat. Cream is a luxury item and may be used as an accompaniment to coffee, as a filling in cakes, and an ingredient in ice cream.

Separation

Separation can very simply be achieved by removing the cream with a spoon, however this is a slow process during which the cream may spoil. For this reason it is more usual to use a manual or powered centrifuge.

Pasteurization

Cream may be pasteurized in a similar way to milk, using a similar time and temperature combination and the same equipment. Cream can also be sterilized but there is a considerable loss of quality.

Packaging and storage

Cream can be packaged in glass jars or plastic pots sealed with foil lids. Pasteurized cream must be stored at a temperature of 4°C to have a shelf-life of several days. Refrigerated storage is necessary because cream is prone to rapid spoilage.

Table 2.2 Production stages for cream

Ingredients	*Process*	*Equipment*	*Section reference*
Raw milk tested	Store at 4°C	Milk churns	62.0
		Refrigerated storage	15.0
		Thermometer	63.0
	Separation of milk fat	Ladle Dairy centrifuge	07.1
	Pasteurization	Large boiling pan or steam jacketed pan	48.0
		Pasteurizer	50.0
		Heat source	36.0
		Thermometer	63.0
	Fill bottles/pots	Funnel or liquid-filling machine	28.1
		Capping machine	47.2
		Pot sealer	47.1
	Cool bottles	Bottle-cooler	See Packaging chapter
	Store bottles at 4°C	Refrigerated storage	15.0

Butter

Butter is a semi-solid mass which contains approximately 80-85 per cent milk-fat and 15-20 per cent water. It is yellow/white in colour, with a bland flavour and a slightly salty taste. It is a valuable product that has a high demand for domestic use in some countries and as an ingredient in other food processing (e.g. for confectionery and bakery uses).

The principles of preservation are:

- to destroy enzymes and micro-organisms by pasteurizing the milk
- to prevent microbial growth during storage by reducing the water content, by storing the product at a low temperature, and optionally by adding a small amount of salt during processing.

Churning

Churning disrupts the emulsion of fat and water and as a result the milk-fat separates out into granules. This process takes place in a butter churn.

Churning cream

Churning is continued until fat granules are present and at this stage the mixture is drained to remove liquid that has separated from the granules. This liquid is known as buttermilk and can be used as either a beverage or as an ingredient in animal feed.

Washing

Clean water equivalent in weight to the buttermilk is added to the churn in order to wash the butter granules. The wash water is drained away. Churning is continued for a short time to compact the butter, and once this has been achieved it is removed from the churn.

Table 2.3 Production stages for butter

Ingredients	*Process*	*Equipment*	*Section reference*
Cream or soured cream	Store at 4°C	Milk churns	62.0
		Refrigerated storage	15.0
		Thermometer	63.0
	Churning	Butter churns	13.0
	Draining (pour off buttermilk)		
	Washing		
	Draining (pour off washwater)		
Permitted colours and	Kneading/working	Butter pats	04.0
salt (optional)	Form into blocks	Butter pats	04.0
	Packaging	Paper/plastic/ foil wrapping	47.3
		Wrapping machines	
	Storage at 4°C	Refrigerated storage	15.0

Forming and packaging

Butter is kneaded to achieve a smooth and pliable texture. This can be done using simple hand-tools such as butter pats. Alternatively for higher production rates a specially-designed kneader can be used. Once the butter has a uniform and smooth texture it is formed into blocks with butter pats and packed in either greaseproof paper or foil wrappers.

Working butter with butter pats

Storage

Due to its high fat composition, butter must be stored at temperatures below 10°C otherwise the fat becomes rancid and imparts undesirable 'off' flavours. The water droplets in butter (20 per cent) can also allow bacteria to grow if it is not kept under cool conditions.

Ghee

Ghee is made from butter which has been heated and clarified. At ambient temperatures it is a semi-solid mass with a granular texture, but on melting (40°C+) it turns into a clear, thin liquid. It has a high demand in some countries for domestic use, as an ingredient for local food production (for example bakeries and confectionery manufacturers), and as an export commodity.

Alternatively, cream is boiled gently to evaporate the water. During boiling the product is stirred continuously until the milk proteins start to coagulate, forming particles, and the colour of the cream darkens. Heating is stopped and the product is left to set. The particles settle at the bottom of the vessel and the milk-fat is separated. The principles of preservation are:

- heating to destroy enzymes and contaminating microorganisms
- to reduce the water-content by evaporation, and in doing so prevent the growth of micro-organisms.

Table 2.4 Production stages for ghee

Ingredients	*Process*	*Equipment*	*Section reference*
Butter	Heating	Heat source	36.0
		Large boiling pan or steam jacketed pan	48.0
	Cool to room temperature	Thermometer	63.0
	Filter	Filter cloth	08.0
	Fill into jars/pots	Funnel or liquid-filling machine	28.1
		Capping machine	47.2
	Store at ambient temperatures		

Packaging and storage

Metal containers are normally used. They should be thoroughly cleaned, especially if they are re-usable, and they should be made airtight. Alternatives to metal cans include coloured glass jars with metal lids, or ceramic pots sealed with cork/plastic stoppers.

Ghee is usually stored at room temperatures as cold storage affects the granular texture. Thus ghee is useful for those consumers with no access to refrigeration.

Cultured/fermented dairy products

The technology of cultured milk products such as yoghurt, curd, and cheese is based upon the microbial conversion of the milk-sugar lactose to lactic acid (lactic acid accounts for the characteristic 'sourness' of such products). In order for the conversion to take place, lactic acid producing bacteria must be present. This may occur by allowing the milk to sour naturally, but it is better to introduce the appropriate bacteria as a starter culture. Starter cultures may be in the form of a small quantity of previously-cultured product or may be purchased as a commercially-prepared culture.

Yoghurt/curd

Yoghurt is a fermented milk product that evolved by allowing naturally-contaminated milk to sour at a warm temperature. Yoghurt can be either unsweetened or sweetened, set, or stirred. Curd is the name given to a yoghurt-type product made from buffalo milk.

The principles of preservation for yoghurt are:

- Pasteurization of the raw milk to destroy contaminating microorganisms and enzymes.
- An increase in acidity due to the production of lactic acid from lactose. This inhibits the growth of food-poisoning bacteria.
- Storage at a low temperature to inhibit the growth of microorganisms.

Table 2.5 Production stages for set yoghurt

Ingredients	*Process*	*Equipment*	*Section reference*
Milk and starter culture (2 per cent)	Preheat to 70°C for 15-20 minutes	Heat source	36.0
		Thermometer	63.0
		Boiling pan	48.0
	Cool to 30-40°C	Thermometer	63.0
	Addition of starter culture	Measuring and weighing equipment	64.1 and 64.2
	Pour into bottles/pots	Funnel or Liquid filler	28.1
		Sealing machine	47.1
		or Capping machine	47.2
	Incubate at 43-45°C	Commercial incubator	39.0
		Thermometer	63.0
	Store at 4°C	Refrigerated storage	15.0

Heating

In the manufacture of yoghurt, milk is normally heated to 70°C for 15-20 minutes, using an open boiling pan, or alternatively a steam jacketed pan.

Addition of starter culture

The milk is cooled to between 30 40°C and inoculated with a mixed culture of *Lactobacillus bulgaricus* and *Streptococcus thermophilus* (usually in a ratio of 1:1). If a commercial starter-culture is used, the directions for use will be given. However, if a culture from a previous batch is used, then it is usual to add 2-3 tablespoons per litre of prepared milk.

Yoghurt of the stirred variety can be fermented in the mixing container. To make set yoghurt the inoculated milk should be poured into the individual pots before fermentation.

Selling curd from a roadside stall

Incubation

The micro-organisms that produce yoghurt are most active within a temperature range of 32-47°C. Ambient temperatures are therefore not adequate and a heated incubator is needed. Small commercially-available yoghurt-makers consist of an electrically-heated base and a set of plastic or glass containers. Most yoghurt-makers make four or five individual half litre cups at a time. There are other simple and inexpensive ways of incubating yoghurt such as an insulated box, keeping the jars/pots surrounded by warm water, or by using thermos flasks (the latter is only suitable for stirred yoghurt). Incubation takes approximately five hours.

When fermentation is complete, stirred yoghurt is cooled and flavoured or sweetened prior to packaging. In set yoghurt all ingredients are added before fermentation.

Packaging and storage

Yoghurt or curd is commonly packaged in plastic pots fitted with a plastic lid, or heat-sealed with foil, although traditionally, curd is packaged in clay pots. Such pots are made from local materials and can be re-used or later used for cooking. The shelf-life of yoghurt is usually 3-8 days when stored at temperatures below 10°C.

Cheese

Cheese is made from milk by the combined action of lactic acid bacteria and the enzyme rennin (known as rennet). Just as cream is a concentrated form of milk fat, cheese is a concentrated form of milk-protein. The differences in cheeses that are produced in different regions result from variations in the composition and type of milk, variations in the process, and the bacteria used. The different cheese varieties can be classified as either hard or soft.

Collecting milk for cheese-making

Hard cheeses such as Cheddar and Edam have most of the whey drained out and are pressed. Soft cheeses such as paneer contain some of the whey and are not pressed. Many indigenous cheeses are soft types.

The hardness, flavour, and other qualities of a cheese can be varied by changes to the process conditions, to suit local tastes. However the principal steps of a cheese-making process are basically the same.

The principles of preservation are:

- the raw milk is pasteurized to destroy most enzymes and contaminating bacteria
- fermentation by lactic-acid bacteria increases the acidity which inhibits the growth of food-poisoning and spoilage bacteria
- the moisture content is lowered and salt is added to inhibit bacterial and mould growth.

The table, right, outlines the stages of production and the equipment needed to produce Edam cheese.

Pre-heating

The pasteurized milk is heated to a temperature at which the starter-culture can work.

Addition of starter culture

Starter-culture is added to the milk at the rate of approximately 2 per cent of the weight of milk. The vessel used should be either aluminium or stainless steel.

Addition of rennet

The rennet should be 1 per cent of the weight of milk. The rennet alters the milk proteins and allows them to form the characteristic curd.

Incubation

The milk is allowed to stand until it sets to a firm curd.

Treatment of the curd

The curd is cut into cubes which facilitate the elimination of whey from the gel. The curd is then cooked at 40°C for a period of twenty minutes which has the action of firming the curd. After cooling, the whey is drained off. The curd is pressed to ensure that most of the whey has been removed, and is then cut to fit the cheese-moulds, and finally pressed with weights.

Ripening

This is the final stage in the cheese-making process. It is a process which allows the development of gas in some cheeses and the development of flavour. The longer the ripening process the stronger the flavour. Ripening usually takes place in ripening rooms, where the temperature and humidity must be controlled for the optimum development of the cheese.

Table 2.6 Production stages for Edam-type cheese

Ingredients	*Process*	*Equipment*	*Section reference*
Pasteurized milk	Preheat to 35-40°C	Cheese vat	10.0
		or boiling pan	48 0
		Thermometer	63.0
		Heat source	36 0
Starter culture	Addition of starter culture	Measuring and weighing equipment	64.1 and 64.2
Rennet	Add rennet at 30°C	Measuring and weighing equipment	64.1 and 64.2
		Thermometer	63.0
	Incubate		
	Cut the curd	Curd cutters	16.1
	Heat to 40°C for 20 minutes	Heat source	36.0
		Thermometer	63.0
	Drain	Filter cloth	08.0
	Cut to fit a cheese	Knife	17.1

Ingredients	Process	Equipment	Section reference
	-mould		
	Put into a cheese -mould	Cheese-moulds	09.1
	Press with	Cheese-press weights	09.2
	Cool and dry at 10-12°C	Thermometer	63.0
Salt	Salting in 20 per cent salt solution at 12°C for 12-16 hours	Brine meter Thermometer	64.6 63.0
	Ripen for 6-8 weeks at 16°C	Thermometer (optional)	63.0.
	Washing		
	Drying for 30 minutes		
	Wax with paraffin wax store at 9°C	Refrigerated storage	15.0

Packaging and storage

The packaging requirements differ according to the type of cheese produced. Hard cheese, for example, has an outer protective rind which protects the cheese from air, microorganisms, light, moisture-loss or pick-up, and odour pickup. Cheese should be allowed to 'breathe', otherwise it will sweat. Suitable wrapping materials are therefore cheesecloth or grease-proof paper. Cheese should be stored at a relatively low temperature between 4 and 10°C to achieve a shelf-life of several weeks/months. Soft cheeses are often stored in pots or other containers, often in brine, to help increase their shelf-life of several days/weeks.

Selling ice cream from a bicycle

Ice cream

Ice cream is a frozen mixture which contains milk, sugar, fat, and optional thickeners (e.g. pectin or gelatin), colouring, and flavouring. It may be sweetened and flavoured in numerous ways with nuts, fruit pieces, and natural or artificial flavours and colours.

The principles of preservation are:

- pasteurization to destroy most micro-organisms and enzymes
- freezing to inhibit microbial growth.

Pasteurization

Pasteurization is carried out by heating to 65°C for a period of 30 minutes.

Cooling and beating

Ice cream is a complex mix of small ice crystals and air bubbles in a milk-fat/water emulsion. To achieve this, it is necessary to cool the mixture quickly to produce small ice crystals and at the same time incorporate air into the product by beating.

Ingredients	*Processing stage*	*Equipment*	*Section reference*
Milk, sugar, fat,	Mix ingredients	Weighing and	64.1 and 64.2
	thickener, colours	measuring equipment	
	and flavours.	Liquid mixer	43.1
	Pasteurize	Boiling pan or steam	48.0
		jacketed pan	
		Thermometer	63.0
		Heat source	36.0
	Cool mixture to approximately	Ice cream maker	38.0
	-5°C and beat simultaneously		
	Fill into containers	Filling machine	28.0
	Freeze at -18°C	Freezer	32.0

Ice cream makers are available commercially and work on the following principle. The mixture is placed in a bowl which is kept at a low temperature (either surrounded by ice and salt, or having been chilled in a freezer). It is then agitated by a manually-operated rotor or by a powered stirrer. At the end of this process the ice cream should be at a temperature of approximately -5°C, and be partly frozen.

Packaging and storage

The ice cream is usually packaged in plastic, waxed paper, or cardboard containers, and is stored at below -18°C. The storage temperature is important for two reasons:

- to maintain the texture of the product
- to prevent the growth of micro-organisms.

Ice cream may be transported in an insulated box (*e.g.* for sale from a bicycle). It is especially important to guard against thawing and re-freezing as this will cause changes in texture and mouthfeel, and there is the increased possibility of food poisoning by contaminating food poisoning microorganisms.

Suitability for small-scale production

In some regions, there is a high demand for dairy products both for traditional and modern items. Much of the technology and machinery necessary for processing is fairly simple, which may at first sight appeal to the would-be processor. Milk, however, is a highly perishable food and there is a high risk of transmitting food-poisoning bacteria to consumers. It is stressed, therefore, that milk processing of any kind must be done under carefully controlled hygienic conditions.

Furthermore, in developing countries, milk processing is often more problematic than in temperate climates, owing to higher temperatures and humidity. Consequently, milk spoils at a faster rate, cheese ripens too quickly, and it is often difficult to ensure adequate cooling conditions.

Ice Cream Manufacture

The basic steps in the manufacturing of ice cream are generally as follows:

- blending of the mix ingredients
- pasteurization
- homogenization
- aging the mix
- freezing
- packaging
- hardening

Blending

First the ingredients are selected based on the desired formulation and the calculation of the recipe from the formulation and the ingredients chosen, then the ingredients are weighed and blended together to produce what is known as the «ice cream mix». Blending requires rapid agitation to incorporate powders, and often high speed blenders are used.

Pasteurization of Mix

The mix is then pasteurized. Pasteurization is the biological control point in the system, designed for the destruction of pathogenic bacteria. In addition to this very important function, pasteurization also reduces the number of spoilage organisms such as psychrotrophs, and helps to hydrate some of the components (proteins, stabilizers).

- Pasteurization (Ontario regulations): 69° C/30 min. 80° C/25s
- Both batch pasteurizers and continuous (HTST) methods are used.

Batch pasteurizers lead to more whey protein denaturation, which some people feel gives a better body to the ice cream. In a batch pasteurization system, blending of the proper ingredient amounts is done in large jacketed vats equipped with some means of heating, usually steam or hot water. The product is then heated in the vat to at least 69°C (155°F) and held for 30 minutes to satisfy legal requirements for pasteurization, necessary for the destruction of pathogenic bacteria. Various time temperature combinations can be used. The heat treatment must be severe enough to ensure destruction of pathogens and to reduce the bacterial count to a maximum of 100,000 per gram. Following pasteurization, the mix is homogenized by means of high pressures and then is passed across some type of heat exchanger (plate or double or triple tube) for the purpose of cooling the mix to refrigerated temperatures (4 °C). Batch tanks are usually operated in tandem so that one is holding while the other is being prepared. Automatic timers and valves ensure the proper holding time has been met.

Continuous pasteurization is usually performed in a high temperature short time (HTST) heat exchanger following blending of ingredients in a large, insulated feed tank. Some preheating, to 30 to 40 °C, is necessary for solubilization of the components. The HTST system is equipped with a heating section, a cooling section, and a regeneration section. Cooling sections of ice cream mix HTST presses are usually larger than milk HTST presses. Due to the preheating of the mix, regeneration is lost and mix entering the cooling section is still quite warm.

Homogenization of Mix

The mix is also **homogenized**, which forms the fat emulsion by breaking down or reducing the size of the fat globules found in milk or cream to less than 1 μ m. Two stage homogenization is usually preferred for ice cream mix. Clumping or clustering of the fat is reduced thereby producing a thinner, more rapidly whipped mix. Melt-down is also improved. Homogenization provides the following functions in ice cream manufacture:

- Reduces size of fat globules
- Increases surface area
- Forms membrane
- makes possible the use of butter, frozen cream, etc.

By helping to form the fat structure, it also has the following indirect effects:

- makes a smoother ice cream
- gives a greater apparent richness and palatability
- better air stability
- increases resistance to melting

Homogenization of the mix should take place at the pasteurizing temperature. The high temperature produces more efficient breaking up of the fat globules at any given pressure and also reduces fat clumping and the tendency to thick, heavy bodied mixes. No one pressure can be recommended that will give satisfactory results under all conditions.

The higher the fat and total solids in the mix, the lower the pressure should be. If a two stage homogenizer is used, a pressure of 2000 - 2500 psi on the first stage and 500 - 1000 psi on the second stage should be satisfactory under most conditions. Two stage homogenization is usually preferred for ice cream mix. Clumping or clustering of the fat is reduced thereby producing a thinner, more rapidly whipped mix. Melt-down is also improved.

Ageing of Mix

The mix is then aged for at least four hours and usually overnight. This allows time for the fat to cool down and crystallize, and for the proteins and polysaccharides to fully hydrate. Aging provides the following functions:

- Improves whipping qualities of mix and body and texture of ice cream
- It does so by:
- providing time for fat crystallization, so the fat can partially coalesce;
- allowing time for full protein and stabilizer hydration and a resulting slight viscosity increase;
- allowing time for membrane rearrangement and protein/emulsifier interaction, as emulsifiers displace proteins from the fat globule surface, which allows for a reduction in stabilization of the fat globules and enhanced partial coalescence.

Aging is performed in insulated or refrigerated storage tanks, silos, etc. Mix temperature should be maintained as low as possible without freezing, at or below 5 C. An aging time of overnight is likely to give best results under average plant conditions. A "green" or unaged mix is usually quickly detected at the freezer.

Freezing/Whipping of Ice Cream

Following mix processing, the mix is drawn into a flavour tank where any liquid flavours, fruit purees, or colours are added. The mix then enters the dynamic freezing process which both freezes a portion of the water and whips air into the frozen mix. The "barrel" freezer is a scraped-surface, tubular heat exchanger, which is jacketed with a boiling refrigerant such as ammonia or freon. Mix is pumped through this freezer and is drawn off the other end in a matter of 30 seconds, (or 10 to 15 minutes in the case of batch freezers) with about 50% of its water frozen. There are rotating blades inside the barrel that keep the ice scraped off the surface of the freezer and also dashers inside the machine which help to whip the mix and incorporate air.

Ice cream contains a considerable quantity of air, up to half of its volume. This gives the product its characteristic lightness. Without air, ice cream would be similar to a frozen ice cube. The air content is termed its overrun, which can be calculated mathematically.

As the ice cream is drawn with about half of its water frozen, particulate matter such as fruits, nuts, candy, cookies, or whatever you like, is added to the semi-frozen slurry which has a consistency similar tosoft-serve ice cream. In fact, almost the only thing which differentiates hard frozen ice cream from soft-serve, is the fact that soft serve is drawn into cones at this point in the process rather than into packages for subsequent hardening.

Hardening

After the particulates have been added, the ice cream is packaged and is placed into a blast freezer at -30° to -40° C where most of the remainder of the water is frozen. Below about -25° C, ice cream is stable for indefinite periods without danger of ice crystal growth; however, above this temperature, ice crystal growth is possible and the rate of crystal growth is dependant upon the temperature of storage. This limits the shelf life of ice cream.

A primer on the theoretical aspects of freezing will help you to fully understand the freezing and recrystallization process.

Hardening invloves static (still, quiescent) freezing of the packaged products in blast freezers. Freezing rate must still be rapid, so freezing techniques involve low temperature (-40 °C) with either enhanced convection (freezing tunnels with forced air fans) or enhanced conduction (plate freezers).

The rate of heat transfer in a frezing porcess is affected by the temperature difference, the surface area exposed and the heat transfer coefficient (Q=U A dT). Thus, the factors affecting hardening are those affecting this rate of heat transfer:

- Temperature of blast freezer - the colder the temperature, the faster the hardening, the smoother the product.
- Rapid circulation of air - increases convective heat transfer.
- Temperature of ice cream when placed in the hardening freezer - the colder the ice cream at draw, the faster the hardening; - must get through packaging operations fast.
- Size of container - exposure of maximum surface area to cold air, especially important to consider shrink wrapped bundles - they become a much larger mass to freeze. Bundling should be done after hardening.
- Composition of ice cream - related to freezing point depression and the temperature required to ensure a significantly high ice phase volume.
- Method of stacking containers or bundles to allow air circulation. Circulation should not be impeded - there should be no 'dead air' spaces (*e.g.*, round vs. square packages).
- Care of evaporator - freedom from frost - acts as insulator.
- Package type, should not impede heat transfer - *e.g.*, styrofoam liner or corrugated cardboard may protect against heat shock after hardening, but reduces heat transfer during freezing so not feasible.
- Ice cream from the dynamic freezing process (continuous freezer) can also be transformed into an array of novely/impulse products through a variety of filling and forming machines, which have ben identified on a separate page.

Chapter 3

Fish and Marine Product Processing

The term fish processing refers to the processes associated with fish and fish products between the time fish are caught or harvested, and the time the final product is delivered to the customer. Although the term refers specifically to fish, in practice it is extended to cover any aquatic organisms harvested for commercial purposes, whether caught in wild fisheries or harvested from aquaculture or fish farming.

Larger fish processing companies often operate their own fishing fleets or farming operations. The products of the fish industry are usually sold to grocery chains or to intermediaries. Fish are highly perishable. A central concern of fish processing is to prevent fish from deteriorating, and this remains an underlying concern during other processing operations.

Fish processing can be subdivided into fish handling, which is the preliminary processing of raw fish, and the manufacture of fish products. Another natural subdivision is into primary processing involved in the filleting and freezing of fresh fish for onward distribution to fresh fish retail and catering outlets, and the secondary processing that produces chilled, frozen and canned products for the retail and catering trades.

There is evidence humans have been processing fish since the early Holocene. These days, fish processing is undertaken by artisan fishermen, on board fishing or fish processing vessels, and at fish processing plants.

Fish is a highly perishable food which needs proper handling and preservation if it is to have a long shelf life and retain a desirable quality and nutritional value. The central concern of fish processing is to prevent fish from deteriorating. The most obvious method for preserving the quality of fish is to keep them alive until they are ready for cooking and eating. For thousands of years, China achieved this through the aquaculture of carp. Other methods used to preserve fish and fish products include

- the control of temperature using ice, refrigeration or freezing
- the control of water activity by drying, salting, smoking or freeze-drying
- the physical control of microbial loads through microwave heating or ionizing irradiation
- the chemical control of microbial loads by adding acids
- oxygen deprivation, such as vacuum packing.

Usually more than one of these methods is used. When chilled or frozen fish or fish products are transported by road, rail, sea or air, the cold chainmust be maintained. This requires insulated containers or transport vehicles and adequate refrigeration. Modern shipping containers can combine refrigeration with a controlled atmosphere.

Fish processing is also concerned with proper waste management and with adding value to fish products. There is an increasing demand for ready to eat fish products, or products that don't need much preparation.

Handling the Catch

When fish are captured or harvested for commercial purposes, they need some preprocessing so they can be delivered to the next part of the marketing chain in a fresh and undamaged condition. This means, for example, that fish caught by a fishing vessel need handling so they can be stored safely until the boat lands the fish on shore. Typical handling processes are transferring the catch from the fishing gear (such as a trawl, net or fishing line) to the fishing vessel holding the catch before further handling sorting and grading bleeding, gutting and washing chilling storing the chilled fish unloading, or landing the fish when the fishing vessel returns to port

The number and order in which these operations are undertaken varies with the fish species and the type of fishing gear used to catch it, as well as how large the fishing vessel is and how long it is at sea, and the nature of the market it is supplying.[3] Catch processing operations can be manual or automated. The equipment and procedures in modern industrial fisheries are designed to reduce the rough handling of fish, heavy manual lifting and unsuitable working positions which might result in injuries.

Handling Live Fish

An alternative and obvious way of keeping fish fresh is to keep them alive until they are delivered to the buyer or ready to be eaten. This is a common practice worldwide. Typically, the fish are placed in a container with clean water, and dead, damaged or sick fish are removed. The water temperature is then lowered and the fish are starved to reduce their metabolic rate. This decreases fouling of water with metabolic products (ammonia, nitrite and carbon dioxide) that become toxic and make it difficult for the fish to extract oxygen.

Fish can be kept alive in floating cages, wells and fish ponds. In aquaculture, holding basins are used where the water is continuously filtered and its temperature and oxygen level are controlled. In China, floating cages are constructed in rivers out of palm woven baskets, while in South America simple fish yards are built in the backwaters of rivers. Live fish can be transported by methods which range from simple artisanal methods where fish are placed in plastic bags with an oxygenated atmosphere, to sophisticated systems which use trucks that filter and recycle the water, and add oxygen and regulate temperature.

Preservation

Preservation techniques are needed to prevent fish spoilage and lengthen shelf life. They are designed to inhibit the activity of spoilage bacteria and the metabolic changes that result in the loss of fish quality. Spoilage bacteria are the specific bacteria that produce the unpleasant odours and flavours associated with spoiled fish. Fish normally host many bacteria that are not spoilage bacteria, and most of the bacteria present on spoiled fish played no role in the spoilage. To flourish, bacteria need the right temperature, sufficient water and oxygen, and surroundings that are not too acidic. Preservation techniques work by interrupting one or more of these needs. Preservation techniques can be classified as follows.

Control of Temperature

If the temperature is decreased, the metabolic activity in the fish from microbial or autolytic processes can be reduced or stopped. This is achieved byrefrigeration where the temperature is dropped to about 0 °C, or freezing where the temperature is dropped below-18°C. On fishing vessels, the fish are refrigerated mechanically by circulating cold air or by packing the fish in boxes with ice. Forage fish, which are often caught in large numbers, are usually chilled with refrigerated or chilled seawater. Once chilled or frozen, the fish need further cooling to maintain the low temperature. There are key issues with fish cold store design and management, such as how large and energy efficient they are, and the way they are insulated and palletized.

An effective method of preserving the freshness of fish is to chill with ice by distributing ice uniformly around the fish. It is a safe cooling method that keeps the fish moist and in an easily stored form suitable for transport. It has become widely used since the development of mechanical refrigeration, which makes ice easy and cheap to produce. Ice is produced in various shapes; crushed ice and ice flakes, plates, tubes and blocks are commonly used to cool fish. Particularly effective is slurry ice, made from micro crystals of ice formed and suspended within a solution of water and a freezing point depressant, such as common salt.

A more recent development is pumpable ice technology. Pumpable ice flows like water, and because it is homogeneous, it cools fish faster than fresh water solid ice methods and eliminates freeze burns. It complies with HACCP and ISO food safety and public health standards, and uses less energy than conventional fresh water solid ice technologies.

Ice manufactured in this ice house is delivered down the Archimedes screw into the ice hold on the boat, Pittenweem

Control of Water Activity

The water activity, a_w, in a fish is defined as the ratio of the water vapour pressure in the flesh of the fish to the vapour pressure of pure water at the same temperature and pressure. It ranges between 0 and 1, and is a parameter that measures how available the water is in the flesh of the fish. Available water is necessary for the microbial and enzymatic reactions involved in spoilage. There are a number of techniques that have been or are used to tie up the available water or remove it by reducing the a_w. Traditionally, techniques such as drying, salting and smoking have been used, and have been used for thousands of years. These techniques can be very simple, for example, by using solar drying. In more recent times, freeze-drying, water binding humectants, and fully automated equipment with temperature and humidity control have been added. Often a combination of these techniques is used.

Physical Control of Microbial Loads

Heat or ionizing irradiation can be used to kill the bacteria that cause decomposition. Heat is applied by cooking, blanching or microwave heating in a manner that pasteurizes or sterilizes fish products. Cooking or pasteurizing does not completely inactivate microorganisms and may need to be followed with refrigeration to preserve fish products and increase their shelf life. Sterilised products are stable at ambient temperatures up to 40°C, but to ensure they remain sterilized they need packaging in metal cans or retortable pouches before the heat treatment.

Chemical Control of Microbial Loads

Microbial growth and proliferation can be inhibited by a technique called biopreservation. Biopreservation is achieved by adding antimicrobials or by increasing the acidity of the fish muscle. Most bacteria stop multiplying when the pH is less than 4.5. Acidity is increased by fermentation, marinationor by directly adding acids (acetic, citric, lactic) to fish products. Lactic acid bacteria produce the antimicrobial nisin which further enhances preservation. Other preservatives include nitrites, sulphites, sorbates, benzoates and essential oils.

Control of the Oxygen Reduction Potential

Spoilage bacteria and lipid oxidation usually need oxygen, so reducing the oxygen around fish can increase shelf life. This is done by controlling ormodifying the atmosphere around the fish, or by vacuum packaging. Controlled or modified atmospheres have specific combinations of oxygen, carbon dioxide and nitrogen, and the method is often combined with refrigeration for more effective fish preservation.

Combined Techniques

Two or more of these techniques are often combined. This can improve preservation and reduce unwanted side effects such as the denaturation of nutrients by severe heat treatments. Common combinations are salting/drying, salting/marinating, salting/smoking, drying/smoking, pasteurization/refrigeration and controlled atmosphere/refrigeration. Other process combinations are currently being developed along the multiple hurdle theory.

Automated Processes

"The search for higher productivity and the increase of labor cost has driven the development of computer vision technology,[12] electronic scales and automatic skinning and filleting machines."

Waste produced during fish processing operations can be solid or liquid.

- **Solid wastes:** include skin, viscera, fish heads and carcasses (fish bones). Solid waste can be recycled in fish meal plants or it can be treated asmunicipal waste.
- **Liquid wastes:** include bloodwater and brine from drained storage tanks, and water discharges from washing and cleaning. This waste may need holding temporarily, and should be disposed of without damage to the environment. How liquid waste should be disposed from fish processing operations depends on the content levels in the waste of solid and organic matter, as well as nitrogen and phosphorus content, and oil and grease content. It also depends on an assessment of parameters such acidity levels, temperature, odour, and biochemical oxygen demand and chemical oxygen demand. The magnitude of waste management issues depends on how much waste volume there is, the nature of the pollutants it carries, the rate at which it is discharged and the capacity of the receiving environment to assimilate the pollutants. Many countries dispose of such liquid wastes through their municipal sewage systems or directly into a waterway. The receiving waterbody should be able to degrade the organic and inorganic waste components in a way that does not damage the aquatic ecosystem.

Treatments can be primary and secondary.

- **Primary treatments:** use physical methods such as flotation, screening, and sedimentation to remove oil and grease and other suspended solids.
- **Secondary treatments:** use biological and physicochemical means. Biological treatments use microorganisms to metabolise the organic

polluting matter into energy and biomass. "These microorganisms can be aerobic or anaerobic. The most used aerobic processes are activated sludge system, aerated lagoons, trickling filters or bacterial beds and the rotating biological contractors. In anaerobic processes, the anaerobic microorganisms digest the organic matter in tanks to produce gases (mainly méthane and CO_2) and biomass. Anaerobic digesters are sometimes heated, using part of the methanc produced, to maintain a temperature of 30 to 35°C. In the physicochemical treatments, also called coagulation-flocculation, a chemical substance is added to the effluent to reduce the surface charges responsible for particle repulsions in a colloidal suspension, thus reducing the forces that keep its particles apart. This reduction in charge causes flocculation (agglomeration) and particles of larger sizes are settled and clarified effluent is obtained. The sludge produced by primary and secondary treatments is further processed in digesting tanks through anaerobic processes or sprayed over land as a fertilizer. In the latter case, care must be exercised to ensure that the sludge is freed of its pathogens."

Quality and Safety

The International Organization for Standardisation, ISO, is the worldwide federation of national standards bodies. ISO defines *quality* as "the totality of features and characteristics of a product or service that bear on its ability to satisfy stated or implied needs."(ISO 8402). The quality of fish and fish products depends on safe and hygienic practices. Outbreaks of fish-borne illnesses are reduced if appropriate practices are followed when handling, manufacturing, refrigerating and transporting fish and fish products. Ensuring standards of quality and safety are high also minimizes the post-harvest losses."

"The fishing industry must ensure that their fish handling, processing and transportation facilities meet requisite standards. Adequate training of both industry and control authority staff must be provided by support institutions, and channels for feedback from consumers established. Ensuring high standards for quality and safety is good economics, minimizing losses that result from spoilage, damage to trade and from illness among consumers."

Fish processing highly involves very strict controls and measurements in order to ensure that all processing stages have been carried out hygienically. Thus, all fish processing companies are highly recommended to join a certain type of food safety system. One of the certifications that are commonly known is the Hazard Analysis Critical Control Points (HACCP).

Hazard Analysis and Critical Control Points

HACCP is a system which identifies hazards and implements measures for their control. It was first developed in 1960 by NASA to ensure food safety for the manned space program. The main objectives of NASA were to prevent food safety problems and control food borne diseases. HACCP has been widely used by food industry since the late 1970 and now it is internationally recognized as the best system for ensuring food safety.

"The Hazard Analysis and Critical Control Points (HACCP) system of assuring food safety and quality has now gained worldwide recognition as the most cost-effective and reliable system available. It is based on the identification of risks, minimizing those risks through the design and layout of the physical environment in which high standards of hygiene can be assured, sets measurable standards and establishes monitoring systems. HACCP also establishes procedures for verifying that the system is working effectively. HACCP is a sufficiently flexible system to be successfully applied at all critical stages -- from harvesting of fish to reaching the consumer. For such a system to work successfully, all stakeholders must cooperate which entails increasing the national capacity for introducing and maintaining HACCP measures. The system's control authority needs to design and implement the system, ensuring that monitoring and corrective measures are put in place."

HACCP is endorsed by the:

FAO (Food and Agriculture Organization)

Codex Alimentarius (a commission of the United Nations)

FDA (US Food and Drug Administration)

European Union

WHO (World Health Organization)

There are seven basic principles:

Principle 1: Conduct a hazard analysis.

Principle 2: After assessing all the processing steps, the Critical control point (CCP) is controlled. CCP are points which determine and control significant hazards in a food manufacturing process.

Principle 3: Set up critical limits in order to ensure that the hazard identified is being controlled effectively.

Principle 4: Establish a system so as to monitor the CCP.

Principle 5: Establish corrective actions where the critical limit has not been met. Appropriate actions need to be taken which can be on a short or long-term basis. All records must be sustained accurately.

Principle 6: Establish authentication procedures so as to confirm if the principles imposed by HACCP documents are being respected effectively and all records are being taken.

Principle 7: Analyze if the HACCP plan are working effectively.

Final Products

Finfish, or parts of finfish, are typically presented physically for marketing in one of the following forms:

- **whole fish:** the fish as it originally came from the water, with no physical processing
- **drawn fish:** a whole fish which has been eviscerated, that is, had its internal organs removed
- **dressed fish:** fish that has been scaled and eviscerated, and is ready to cook.
- **pan dressed fish:** a dressed fish which has had its head, tail, and fins removed, so it will fit in a pan.
- **filleted fish:** the "fleshy sides of the fish, cut lengthwise from the fish along the backbone. They are usually boneless, although in some fish small bones called "pins" may be present; skin may be present on one side, too. Butterfly fillets may be available. This refers to two fillets held together by the uncut flesh and skin of the belly"
- **fish steaks:** large dressed fish can be cut into cross section slices, usually half to one inch thick, and usually with a cross section of the backbone
- **fish sticks:** "are pieces of fish cut from blocks of frozen fillets into portions at least 3/8-inch thick. Sticks are available in fried form ready to heat or frozen raw, coated with batter and breaded, ready to be cooked"
- **fish cakes:** are "prepared from flaked fish, potatoes, and seasonings, and shaped into cakes, coated with batter, breaded, and then packaged and frozen, ready-to-be-cooked"
- fish fingers
- fish roe

Value Addition

Fish Oil Capsules

In general value addition means "any additional activity that in one way or the other change the nature of a product thus adding to its value at the time of sale." Value addition is an expanding sector in the food processing industry, especially in export markets. Value is added to fish and fishery products depending on the requirement of different markets. Globally a transition period is taking place where cooked products are replacing traditional raw products in consumer preference.

"In addition to preservation, fish can be industrially processed into a wide array of products to increase their economic value and allow the fishing industry and exporting countries to reap the full benefits of their aquatic resources. In addition, value processes generate further employment and hard currency earnings. This is more important nowadays because of societal changes that have led to the development of outdoor catering, convenience products and food services requiring fish products ready to eat or requiring little preparation before serving."

"However, despite the availability of technology, careful consideration should be given to the economic feasibility aspects, including distribution, marketing, quality assurance and trade barriers, before embarking on a value addition fish process."

- **Surimi:** Surimi and surimi-based products are an example of value added products. Surimi is prepared from the mechanically deboned, washed (bleached) and stabilised flesh of fish. "It is an intermediate product used in the preparation of a variety of ready to eat seafood such as kamaboko, fish sausage, crab legs and imitation shrimp products. Surimi-based products are gaining more prominence worldwide, because of the emergence of Japanese restaurants and culinary traditions in North America, Europe and elsewhere. Ideally, surimi should be made from low-value, white fish with excellent gelling ability and which are abundant and available year-round. At present, Alaskan pollack accounts for a large proportion of the surimi supply. Other species, such as sardine, mackerel, barracuda, striped mullet have been successfully used for surimi production."
- **Fishmeal and fish oil:** "A significant proportion of the world catch (20 percent) is processed into fishmeal and fish oil. Fishmeal is a ground solid product that is obtained by removing most of the water and some or all of the oil from fish or fish waste. This industry was launched in the 19th century, based mainly on surplus catches of herring from seasonal coastal fisheries to produce oil for industrial uses in leather tanning and in the production of soap, glycerol and other non-food products. Presently, it uses small oily fish to produce fishmeal and oil. It is worthy to mention that, only where it is uneconomic or impracticable for human consumption, should the catch be reduced to fishmeal and oil. Indeed, cycling fish through poultry or pigs is a loss because there is a need for 3 kg of edible fish to produce approximately 1 kg of edible chicken or pork."

Fish Products

Fish and fish products are consumed as food all over the world. With other seafoods, it provides the world's prime source of high-quality protein: 14–16 percent of the animal protein consumed worldwide. Over one billion people rely on fish as their primary source of animal protein.

Fish and other aquatic organisms are also processed into various food and non-food products.

In Ancient Roman society, garum, a type of fish sauce condiment, was popular.

Sharkskin and rayskin which are covered with, in effect, tiny teeth (dermal denticles) were formerly used in the same manner as sandpaper is in the modern era. These skins are also used to make leather. Rayskin leather (same'gawa) is used in the manufacture of hilts of traditional Japanese swords. Some other species of fish are also used to make fish leather, and this material is more and more popular among luxury brands such as Prada, Dior, Fendi, and also emerging designers.

Thus, it is now possible to wear shoes made of salmon leather, a jacket made of perchleather, or a handbag made of wolffish or cod leather. Once tanned, the leather is non-odorous and is stronger than other, traditional, leathers of similar thickness.

The flesh of many fish are primarily valued as a source of food; there are many edible species of fish, and many fish produce edible roe. Other marine life taken as food includes shellfish,crustaceans, and sea cucumber. Sea plants such as kombu are used in some regional cuisine.

Processed Fish Products

Surimi refers to a Japanese food product intended to mimic the meat of lobster, crab, and other shellfish. It is typically made from white-fleshed fish (such as pollock or hake) that has been pulverized to a paste and attains a rubbery texture when cooked.

Fish glue is made by boiling the skin, bones and swim bladders of fish. Fish glue has long been valued for its use in all manner of products from illuminated manuscripts to the Mongolian war bow.

Fish oil is recommended for a healthy diet because it contains the omega-3 fatty acids, eicosapentaenoic acid (EPA), and docosahexaenoic acid (DHA), precursors to eicosanoids that reduceinflammation throughout the body.

Fish emulsion is a fertilizer emulsion that is produced from the fluid remains of fish processed for fish oil and fish meal industrially.

Fish hydrolysate is ground up fish carcasses. After the usable portions are removed for human consumption, the remaining fish body – guts, bones, cartilage, scales, meat, etc. – are put into water and ground up.

Fish meal is made from both whole fish and the bones and offal from processed fish. It is a brown powder or cake obtained by rendering pressing the whole fish or fish trimmings to remove the fish oil. It used as a high-protein supplement in aquaculture feed.

Fish sauce is a condiment that is derived from fish that have been allowed to ferment. It is an essential ingredient in many curries and sauces.

Isinglass is a substance obtained from the swim bladders of fish (especially sturgeon), it is used for the clarification of wine and beer.

Tatami iwashi is a Japanese processed food product made from baby sardines laid out and dried while entwined in a single layer to form a large mat-like sheet.

Other Processed Products

- Pearls, mother-of-pearl, and abalone are valued for their lustre. Traditional methods of pearl hunting are now virtually extinct.
- Sea horse, star fish, sea urchin and sea cucumber are used in traditional Chinese medicine.

- The Sea snails *Murex brandaris* and *Murex trunculus* are used to make the pigment Tyrian purple.
- Some sepia pigment is made from the inky secretions of cuttlefish.
- Kelp is a major source of iodine, can be used as fertilizer, and kelp ash can be used in soap and glass productions.

Byproducts

A shimmery substance found on fish scales, most usually obtained from herring and one of many by-products of commercial fish processing, can also be used for pearlescent effects, primarily in nail polish, but is now rarely used due to its high cost, bismuth oxychloride flakes being used as a substitute instead.

Fish Protein Powder

Fish protein powder (FPP) describes a food grade powder product designated primarily for human consumption applications. It differs significantly from fish meal products which are designated for animal feed applications. Fish protein powders have various sanitary processing, purity and functional characteristics which establish them as human food ingredients. Production plants registered for the USA market are located in Peru and France.

Historically, the fish processing methods used for human consumption have been: fresh, canned, frozen, smoked or dehydrated - all of which would be used as a whole food rather than as an ingredient in other foods. Additionally, an industrial fish industry exists where whole fish and by products from fish processing have been cooked and dehydrated to form a product termed fish meal, which is used for animal feed, pet food and fish feed.

With the evolution of refining and processing technology and expanded research on the nutrition of fish proteins and peptides, a new industry has developed for the specific purpose of producing a fish protein powder for human consumption with the intent of reaching new ingredient uses and markets. The FPP end product is now used in a variety of food ingredient applications including sports nutrition, food additives and supplements, all of which depend on the finished fish protein powder produced such that it is hygienically safe and also meets sensory requirements of taste, odor and function in prepared foods.

Process

The process of producing fish protein powder starts with the decision of how to prepare, separate and refine the proteins. The raw material can be whole fish, fillets or skins or any combination pending the desired finished product. In order to reach a final powder, an initial step of hydrolyzing the proteins chains to break them down into smaller particles is used. Methods may include one or more of the following: acid or alkaline hydrolysis, thermal hydrolysis or enzymatic hydrolysis. Enzymatic hydrolysis similar to the body's natural digestive process provides the most efficient breakdown of the proteins into smaller fractions termed peptides which can then be separated from the oil and non-digested proteins during liquid phase processing. Subsequent steps of solids and oil removal through various mechanical separation techniques are required to create a final fish protein fraction with acceptable organoleptic properties for use in human food. Minimization of odor through the elimination of fat and oil from the protein fraction, as well as separating out the lowest molecular weight protein fractions from the larger fractions all serve to create a refined fish protein. Some processes utilize solvents to extract the fat but these can result in dangerous handling and potential residual issues. The final step in producing the product is typically spray drying, which involves atomizing the liquid protein in a hot air chamber resulting in rapid evaporation of the water and a fine powder falling to the bottom of the chamber for removal. Spray drying is different from typical oven dehydration used in animal grade fish meal production in that spray drying will not denature the protein fraction resulting in a higher quality protein product for human consumption.

Categories

The two basic categories used to classify fish protein powders are dependent on the levels of protein, fat, mineral and carbohydrate contained in the powder. The minerals are mostly naturally occurring, organic complexes of magnesium, calcium and phosphorus. The spray drying process may utilize other minerals and carbohydrates to improve flow characteristics of the final product thus altering the natural balance. Powders will all have a residual moisture content in the 4-8% range.

- Fish protein concentrate (FPC) - is a powder product with medium level of protein (50-70%) and will contain some level of fat/oil (1-20%) in the powder form as well.
- Fish protein isolate (FPi) - where the product contains less than 1% fat/oil and more than 90% protein.
- New manufacturing techniques are also producing *hybrid FPi* products where the fat/oil content is very low, (<0.3%) with the protein levels in the 80% range. The hybrid FPi does not reach 90% protein (often a definition point for an isolate) as the natural minerals are not removed and thus represent up to 15% of the final mass balance.

Peptides vs proteins and amino acids in the digestive tract

Any animal that consumes a whole protein must break down and digest the protein order to absorb the nutrients. For humans this begins with chewing and the addition of saliva enzymes, followed by acid and protease enzyme digestion in the stomach, whereby the end result is a peptide or amino acid fraction ready for uptake into the blood stream via the small intestine. Research has confirmed that most animals have more Peptide receptors in the gut and lower intestine that they do free amino acid receptors - as such the peptide form of fish protein powder is most conducive for optimal nutritional benefits.

Hygienic production of fish protein powder mimics these natural digestion steps, and pending the degree of hydrolysis, the protein powder will actually be a partial or complete peptide powder, ready for immediate absorption in the intestine.

Nutritional aspects

Significant elements of the nutritional science of fish protein powders centers around the bioactive and antioxidant properties of the peptide fractions produced during hydrolysis and their ability to have a positive impact on many conditions including gastrointestinal issues associated with irritable bowel syndrome (IBS) and Crohn's disease - as well reduction affects onhypertension and fast absorption functionality promotes the addition of lean muscle mass to humans consuming the products. Further studies showed that peptides in fish protein powders can minimize injurious affect of anti-inflammatory pain drugs. The University of Maryland School of Medicine concluded that certain peptide fractions from fish may inhibit prostate cancer and possibly other cancers from spreading.

Additional benefits of fish protein powders are centered around diet needs of various subsets of the human population. Individuals who have lactose intolerance, milk allergy, gluten intolerance orcoeliac disease (aka Celiac's) require alternate protein sources. The low molecular weight profile of hydrolyzed fish protein powders (peptides) also mean they can be used in hypoallergenicapplications since they are no longer full length proteins, but are peptide fractions so the immune response to the original full length protein will not be triggered in many individuals.

The minerals [Mg, Ca] are naturally complexed with amino acids making them more absorbent in the body than inorganic mineral complexes found in many supplements. Additionally, the Mg to Ca ratio is typically greater than 1:1 which is nutritionally significant since Mg is critical in the body's absorption of calcium, and Mg deficiency is now understood as a significant issue in western diets.

The purpose of this document is to explain the complexity of the industry and to provide information on the planning and operation of a fishmeal and oil industry. The production of meal and oil is a sophisticated process requiring considerable skill and experience. The products will meet commercial and international standards only if they are manufactured under hygienic conditions in properly operated and controlled plants.

Raw Material

1. Resources

Fish used for reduction to meal and oil may be divided into three categories:

- fish caught for the sole purpose of fishmeal production (for example by Chile, Peru, Norway, Denmark, South Africa and the USA);
- by-catches from another fishery (by most fish-producing countries);
- fish offcuts and offal from the consumption industry. (The UK and Germany use these materials to produce white fish meal; South Africa

makes rock lobster meal from the carapaces and other parts which are not utilized).

A fishmeal industry requires a regular supply of raw material. When planning fishmeal factories, it is necessary to know the type of fish species available, the length of fishing season, the location of the fish, the catchability of the fish by different fishing gear and, if possible, the attainable catches per year for a continuous period.

2. Composition of the Fish

Practically all fish species as well as most other marine animal life may, in principle, be converted into fish meal. A wide variety of fish species is used for the production of fish meal and oil in different countries. Table 1 gives some examples of constituents of whole fish from different stocks. The composition and quality of the raw material are predominant factors in determining the properties and yield of the products. A separation of fatty substances (lipids) from the other constituents of fatty marine animals is one of the major operations in the manufacture of fish meal and oil.

Gadoids (the cod-like fishes) comprise a number of fish species which can be classified as lean. It is characteristic of these species that most of their fat is located in the liver. Fish meal made from these lean species of fish is called white fish meal.

Clupeids (the herrings) provide the largest single source of raw material for production of fish meal and oil. They may be classified as fatty although the fat content may vary from 2% to 30%, depending on species and season. The fat is not, as in lean fish, concentrated in the liver, but is generally distributed throughout the body.

Scombroids (the mackerels) are also fatty fish species.

Elasmobranchs (the sharks and the rays) are not specially caught for meal and oil production. Some species, however, provide raw material as trash fish and as offal from processing.

Salmonoids (the salmons and other closely related fish) are generally not harvested for fishmeal production, but offal from salmon is used. However, there is one species, the capelin, that has become a considerable source of material for meal and oil.

Crustaceans. The carapaces and shells are used, as well as small crustaceans that are unmarketable for direct human consumption.

3. Sampling of Raw Materials

As the composition of the fish may vary widely during the year, systematic sampling and analysis of seasonal variations provide important information when considering the establishment of a fishmeal industry.

4. Economic Evaluation of the Raw Material

From reliable analyses of the raw material one may estimate the amount of fish meal and oil which may be produced, and hence the value of the raw material. The water content gives basic figures for the cost of drying. There is an interrelationship between fat and water in fish: fat and water are complementary constituents inasmuch as fat replaces water in the flesh due to seasonal variation. For a given quantity of raw material, increasing fat content will lead to improved oil yield, reduced demand on drying energy and increased processing capacity of the plant.

In some countries where the fishing fleet and the fishmeal industry are separate commercial entities, the composition and extent of deterioration of the raw material are often used for evaluating the price. Where the fishing fleet is operated as an integral part of the industry, such evaluation may be used to predict product yield and quality and as a control of production efficiency.

Evaluation requires that representative samples from each catch be submitted for analyses at a control laboratory. Sampling is not straightforward because considerable variations in fish size and quality may exist within the same catch. In purse seines, for example, there appears to be a tendency for the smaller fish to sink to the bottom while the larger fish mass in the upper part of the nets. When the catch is transferred, there is therefore a corresponding segregation of fish; consequently the collector must exercise care to obtain samples representative of the whole catch.

Most conveniently the sampling is done during the unloading of the fishing vessel. An automatic sampling device, installed immediately after the weighing equipment, is recommended, but sampling might be done by hand. Stratified spot sampling of 4-5 kg samples of fish should be made no less than 30 times during unloading of a vessel containing industrial fish. The sampling container will finally hold about 250 kg fish which has to be ground and thoroughly mixed before further sub-dividing of the sample can take place. The final sample of fish for laboratory examination will be approximately 500 g.

5. Methods of analysis

The evaluation of raw material requires the carrying out of a number of proximate methods of analysis, mainly to determine protein, fat, water and ash contents, and the determination of volatile basic nitrogen (usually expressed as mg^{-N}/l00 g minced fish) (see Section 2.6.3). Under practical conditions it is often sufficient to make determinations only of water and volatile basic nitrogen in the samples.

6. Deterioration of the Raw Fish

The rapid deterioration of fish is due to the action of bacteria from the surfaces and the digestive tract and to autolytic breakdown caused by enzymatic action in the tissues and in the tract. Bacterial and autolytic deterioration result in breakdown of both the lipid and protein fractions.

Autolytic deterioration

In a number of fish species used for fishmeal production, particularly small pelagic fish species such as sardines, anchovies and herring, the digestive enzymes may cause extensive autolysis leading to softening of the meat, rupture of the belly wall and formation of considerable amounts of blood water containing both protein and oil. This process is aggravated by the production of large quantities of gastric enzymes at the time of feeding. Such solubilization causes difficulties in handling and processing and may lead to serious losses of both protein and oil.

Fat deterioration (lipolysis) caused by different fat splitting enzymes (lipases) is a general feature in fatty fish. Fish oils are largely composed of glycerol combined with fatty acids to form glycerides. Splitting of the glycerides of the oil and formation of free fatty acids (FFA) result in reduced quality of the oil with economic consequences.

Oxidation

Oxidation of lipids (rancidity) and browning of the oil occurs under aerobic (in the presence of oxygen) storage conditions; but in transport vessels and storage bins the condition in the interior of the mass of fish is anaerobic (oxygen is absent).

Microbial spoilage

The anaerobic conditions of bulk storage of whole fish create a complex medium in which microbes can grow, with formation of a variety of chemical spoilage products. Some important end products are the volatile basic nitrogenous compounds (mainly ammonia and trimethylamine) and the amount of total volatile basic nitrogen (TVB-N) is often used as a measure of deterioration. Some volatile bases are formed by the bacterial breakdown of amino acids, in turn derived from protein, but the trimethylamine is formed by bacterial metabolism of trimethylamine oxide. The extensive production of ammonia in a load of deteriorating fish may result in significant losses of protein. In elasmobranchs ammonia is also produced from the urea which is a constituent of their blood and muscle.

The chemical compounds resulting from bacterial activity are numerous and some are not yet well described, but the sulphur containing compounds, which seem to be formed mainly under anaerobic conditions, are significant. Hydrogen sulphide (H_2S) and mercaptans may be produced by decomposed fish in lethal concentrations in the holds of fishing vessels and in closed fish bins. In Denmark, for example, thorough ventilation is compulsory before and during the unloading of fish.

Another feature of bacterial deterioration is the transformation of sulphur from sulphur containing amino acids into compounds which are catalyst inhibitors; that is they inactivate the catalyst used in subsequent hydrogenation of the oil into fat for margarine production.

During meal production, the aqueous phase is particularly vulnerable to bacterial putrefaction. Bacterial breakdown, besides affecting the yield and quality of the end products and the production capacity, also results in the formation of malodorous compounds. Consequently efforts should be made to minimize bacterial spoilage.

7. Preservation of the Raw Material

The production of fish meal and oil from fresh raw material gives the highest yield and the best quality final products.

In many cases, however, it is difficult to avoid partial spoilage as the fish has to be collected from remote areas. The search for economic means of preserving the catch during periods of transport and storage, exceeding about 30 h, has been a continuing challenge to the industry. As breakdown of fish protein and oil are due to, both autolytic and microbial activity, the preserving method should preferably retard both bacterial growth and autolysis by digestive and tissue enzymes. The storage life of fish can be extended either by physical or chemical means.

Draining

Proper draining of the fish, both aboard and ashore, is a simple and effective method of extending the short-term storage life of fish. Aboard the fishing vessels proper drainage reduces the amount of rubbing together and breakage of the fish during rolling and pitching of the fishing vessel. Moreover, the spreading and rapid growth of bacteria is reduced by restricting the presence of free water containing body slime, gut contents and the bacteria contained therein.

Preservation by chilling

During storage of ungutted fish in bulk, the rate of breakdown caused by bacteria and by digestive and tissue enzymes doubles with an increase of temperature of approximately 4°C. This breakdown leads to losses of protein and oil and reduces the quality of the fish for processing. At temperature higher than 5°C hydrogen sulphide is produced by bacteria, and its formation rapidly increases with temperature. At 0°C hydrogen sulphide is not formed until storage exceeds 9-10 days (Petersen, 1971).

In tropical and temperate areas where the fish may be caught at high temperatures and far from the plant, chilling is the most effective method of preserving bulk-stored fish. In many cases the cost of chilling can be recovered by the reduced losses of protein and oil.

In principle, two methods of chilling may be considered, namely refrigerated or chilled water systems and mixing of ice with fish.

Refrigerated (chilled) water systems

Sea water is readily available, but prolonged storage of fish in sea water is limited by absorption of salt; high salt levels are undesired in fish meal. It is therefore preferable to use diluted sea water or fresh water. The method comprises circulating the refrigerated water through the mass of fish. Merely pumping the water over the fish does not effectively chill the whole mass, as there is hardly any penetration of the water through the bulk of the fish, most of it being diverted around the sides. Advanced systems are based on pumping the chilled water upward through the fish from the bottom of the hold. The method is expensive and hardly practical except for long voyages and storage periods.

Preservation by ice

Mixing fish and ice in the right proportion for chilling the fish at 0°C is an effective method for preserving raw fish. Fish and ice should be mixed before filling the hold. The melt water is drained off from the bottom leaving the fish dry and compact. The use of ice for preservation is dependent upon the development of rapid, preferably automatic, systems for mixing fish and ice at the high rate necessary in industrial fishing.

In Scandinavia, particularly in Denmark, the following ice mixing (and grading) system has been installed in more than 100 fishing vessels (see Figure l).

The Technological Laboratory of the Danish Ministry of Fisheries has participated in the development of a receiving box on deck equipped with a conveyor continuously feeding a rotating drum. This drum grades the fish into two categories: industrial, which includes fish less than 35 mm thick, and food fish, which are comprised of thicker specimens.

The receiving box is equipped to remove the few large fish and other large objects before the catch enters the conveyor and the grader. The box may receive loads of up to 2 t of fish, which are converted to a continuous flow of up to 1 200 kg/min. The food fish pass through the length of the cylinder, while the industrial fish fall through the grader into a trough equipped with a continuous supply of small pieces of ice to chill the fish to 0°C and maintain this temperature until the fish are landed. A conveyor running along the trough takes the mixture of industrial fish and ice to a vertical conveyor that lifts the mixture 2 m above the deck and releases it into a funnel. Wide plastic tubes connect the funnel to the ice scuttle on the deck over the section of the hold that is to be filled with iced industrial fish.

The continuous supply of ice to the trough under the grader is taken from a. store of bulk ice. At the bottom of the store a horizontal conveyor feeds the ice onto a vertical conveyor to the trough. The conveyor speed is adjustable so that the supply of ice can be varied. At landing, there should be little or no surplus ice. Fish at 15°C when caught requires about 23% of its weight in ice to be chilled and maintained at 0°C for four days in an insulated hold.

Chemical preservation

Chemical preservatives for raw fish immediately act on the bacteria on the surface of the fish, but there is some delay in action on the interior (stomach and intestines) depending upon the rate of penetration of the preservative.

Sodium nitrite, sodium sulphite, ascorbic acid, benzoic acid and many more preservatives have been evaluated, but they are used only to a very limited extent. Sodium nitrite has demonstrated comparatively favourable properties for preservation of species such as herring, as it retards significantly the development of spoilage microbes and reduces the formation of free fatty acids, but unless nitrites are added in strictly controlled small amounts (as is the case in Norway) they may react with other components of the raw material and form nitrosamines which are harmful carcinogenic chemical compounds.

Special measures must be taken when meal is manufactured from fish which has been preserved by nitrite. The use of nitrite must therefore be permitted only under the most careful supervision and should not be encouraged.

Formaldehyde is widely used and, under certain circumstances, also exerts a beneficial firming effect on the raw material so that it is rendered more amenable to pressing, after cooking, for removal of oil. Formaldehyde combines with protein by a mechanism akin to tanning and the reactive amino acid lysine is involved. The use of small amounts, however, (for example 0.05% formalin based on the weight of the fish) has no detectable deleterious effect on protein quality.

- **The Process**
    - To understand the principles of fishmeal and oil manufacture, it is necessary to consider the raw material as composed of three major fractions: solids (fat-free dry matter), oil and water. The purpose of the process is to separate these fractions from each other as completely as possible, with the least possible expense and under conditions rendering the best possible products.
    - Fish can be reduced to meal and oil in a number of ways. Common to all methods of practical importance are the following processing steps:
    - heating, which coagulates the protein, ruptures the fat depots and liberates oil and physico-chemically bound water;
    - pressing (or occasional centrifugation), which removes a large fraction of the liquids from the mass;
    - separation of the liquid into oil and water (stickwater). This step may be omitted if the oil content of the fish is less than 3%;
    - evaporation of the stickwater into a concentrate (fish solubles);
    - drying of the solid material (presscake) plus added solubles, which removes sufficient water from the wet material to form a stable meal,
    - grinding the dried material to the desired particle size.

This chapter deals in some detail with the most generally practised method of large-scale production. A more cursory description is given of small-scale processing and plants designed to handle small and irregular landings of fish or to yield products with special properties. Although the basic principles of the process may appear few and simple, a variety of elements is required to make a smooth running and profitable factory. There are also different types of equipment that will do the same job but in different ways. The number of options have increased considerably during the last years, particularly in the fields of energy saving, automation and environmental protection. The prospective manufacturer should therefore devote ample time and attention for consultations with the most competent suppliers of machinery and complete factories, before making his final decisions on layout and equipment.

Here are some factors that need special consideration at the planning stage. The most important prerequisite for a profitable fishmeal project is an ample and regular supply of raw material at an acceptable price. Length of season, that is how many days per year the plant may be in operation, is of the utmost importance for the profitability of the venture. Furthermore, the longer the season, the greater the weight that should be placed on measures to reduce the variable costs, and the more that should be spent on investments to save labour and energy and to ensure higher yields and quality of the products.

Another important point relates to the size of plant needed for the amounts of raw material in question. Capacities given by the equipment manufacturer, particularly for the cooker and press installation, may serve only as indicators at the preliminary discussions. The yield is an empirical value entirely dependent upon the nature of the raw material to be processed, its size, freshness and texture. If experience is not available for the actual fish species it is, therefore, advisable to run some tests to evaluate the behaviour of the raw material and the performance of the cooking/pressing operation.

The composition of the raw material expressed in terms of dry matter, protein and fat, will determine what yield of products may be expected. The fat content is of special importance at the planning stage, because it will decide whether it will pay to install equipment for the recovery of oil, besides telling what yield of oil may be expected. For all practical purposes, one may reckon that for the processing of fish with less than 3% of fat, one may omit the installation of oil-recovery equipment.

Evaporators for the stickwater, however expensive, may today be considered standard items of equipment for fishmeal plants, because they recover dry matter that will increase the yield of meal by 20% or more, depending upon the freshness and nature of the raw material. Furthermore, they eliminate a serious pollution problem that today appears unacceptable, particularly in the vicinity of densely populated areas. For small factories and irregular operation, omission of evaporators may sometimes be justified, but a final decision on this point should be taken only after careful calculations have been made and the pros and cons thoroughly evaluated.

Location of the plant relative to habitation and closed harbours is another feature of the utmost importance for trouble-free operation. Public regulations protecting the environment against undesirable pollution of air and water should be carefully studied, as these will affect the selection of equipment and factory premises and determine the amount necessary for investment in unproductive installations for odour abatement and cleaning of waste water emissions.

1. The Principal Method of Processing

The bulk of the world's fish meal and oil is today manufactured by the wet pressing method. The main steps of the process are cooking for coagulation of the protein thereby liberating bound water and oil, separation by pressing of the coagulate yielding a solid phase (presscake) containing 60-80% of the oil-free dry matter (protein, bones) and oil, and a liquid phase (press liquor) containing water and the rest of the solids (oil, dissolved and suspended protein, vitamins and minerals). The main part of the sludge in the press liquor is removed by centrifugation in a decanter and the oil is subsequently removed by centrifuge. The stickwater is concentrated in multi-effect evaporators and the concentrate is thoroughly mixed with the presscake, which is then dehydrated usually by two-stage drying. The dried material is milled and stored in bags or in bulk. The oil is stored in tanks.

The raw material is first unloaded from the fishing vessel by crane, wet fish pump, pneumatic elevator or some sort of mechanical conveyor. The fish is weighed or measured by volume before it is transported to the pits or tanks for the storage of raw material.

Large fish are hashed (A) while smaller fish (for example, those less than 40 cm long) are fed directly at a constant rate by the feeding machine (B) to the indirect steam cooker (C). The coagulated mass is pre-strained in a strainer conveyor (D), or in a vibrating screen, before entering the twin screw press (E). The products from the press (presscake and press liquor) are treated as follows. The presscake is disintegrated in the tearing machine (wet mill)(F) to facilitate mixing with stickwater concentrate (see below) and drying in an indirect steam dryer or a direct flame dryer (G). The meal passes through a vibrating screen (H) furnished with a magnet to remove extraneous matter like pieces of wood and metal (for example, fish hooks) before entering the hammer mill (J). The ground meal is automatically weighed out in bags by the scales (K), the bags are closed (e.g., by sewing) and conveyed to the store. Alternatively, the meal is stored in a holding and blending silo before bagging, pelleting or storing in bulk.

To remove most of the sludge, the press liquor passes through a decanter (N). The press liquor then passes through a buffer tank (O) before separation into oil, stickwater and fine sludge in the stickwater centrifuge (P).

The sludge is added to the presscake. The oil passes through a buffer tank (R) before water f and sludge impurities are removed (polishing) in the oil separator (S). After polishing, the oil often passes through an inspection tank before storage in the oil tank. The stickwater passes through a buffer tank (T) before concentration in a multi-effect evaporator (U). After the buffer tank (V), the concentrate is mixed thoroughly with decanter-sludge and presscake before drying. In some cases the stickwater concentrate, called condensed fish solubles, is sold separately.

The factory can be deodorized by air suction from all tanks and machinery. The air passes through a scrubber {L) and is then burned in the steam boiler or treated with chlorine, after which a further scrubber removes residual chlorine.

As stated, large fish have to be hashed into smaller pieces before being passed by the feeder into the cooker. This is to ensure uniform processing and equal temperature in the cooked material. Also, the feeder ensures a steady rate of presentation to the cooker. Figure 3 illustrates one type of hasher often used. It consists of a rotor with staggered knives and a frame with a row of stationary knives.

Heating ("cooking")

The purpose of the heating process is to liberate the oil from the fat depots of the fish, and to condition the material for the subsequent treatment in the various processing units of the plant. "Cooking", as this operation has traditionally been called, is therefore a key process of the utmost importance for the whole functioning of the factory.

Until fairly recently, the general view has been that the best results and optimum performance of the plant would be obtained at the highest possible temperature which, at atmospheric pressure, would be 100 °C. New experiments, however, have shown that the walls of the fat cells are broken down before the temperature reaches 50 °C. The oil is then free, and theoretically it should be possible to separate it from the solid material. Another important observation from recent investigations is that coagulation of the fish protein is completed at about 75 °C and, furthermore, that the process is very rapid. This new experience leads to the conclusion that there is very little, if anything, to be gained by heating the material beyond 75°C or by using a long heating time. The problem is primarily a question of heat transfer and temperature control to ensure a uniform, optimum temperature throughout the whole mass. Since reduction of heat load on the material, that is the combined effect of temperature and time, tends to improve the quality of the products, we may expect new technological answers to the heating problem, in line with this new knowledge. However, at the present state of technology, we have to accept that optimum conditions for a particular type of raw material must largely be established through practical experience.

The most common practice is to cook the fish in a steam cooker, through which it is conveyed continuously. Heat is generally transferred indirectly from a surrounding jacket and a heated rotary screw conveyor. This is an improvement over the direct steam injection cooker, in which water is condensed in the mass during the process and has to be removed by the press and then evaporated from the press liquor. However, in indirect cookers provision is also made for the admission of live steam directly into the mass as this may sometimes be advantageous.

Cooking is an exacting operation in production and is sometimes difficult to control. Production of cooked material which can be readily pressed is dependent on the quality of the raw material and on the process conditions. A precise time-temperature programme for this process can therefore not be set up and, as mentioned above, a process of trial and error is generally required when fish of unknown history is processed. The most common practice of cooking good raw material, however, is to heat to 95°-100 °C within 15 to 20 min. Most manufacturers operate cookers to ensure rapid heating of the mass to a temperature of about 95° C. The proof of good cooking is good pressability of the mass which leads to proper removal of press liquor and, in particular for fatty fish species, efficient recovery of oil, giving a meal with low fat content which is a criterion of quality. The process must be controlled to ensure sufficient cooking, but overcooking must be avoided as this results in problems with pressing and the presence of large amounts of suspended particles in the stickwater, which makes evaporation difficult.

A typical continuous indirect cooker is shown in Figure 6. The cooker is designed as a cylinder having a steam heated jacket throughout and a steam heated rotor, designed as a screw conveyor with hollow flights. The cooker is equipped with covers throughout for inspection and cleaning and with a nozzle system for blowing direct steam into the mass. The cooker may be provided with automatic temperature control equipment, automatic level control for raw material feeding, discharge control equipment (which is required particularly for handling soft raw material) and a trap for collecting heavy foreign matter like stones and scrap iron.

Cookers like this are generally available in sizes which can process from 16 t to 1 600 t of raw material per 24 h.

The capacity of a heat exchanger, such as an indirect steam cooker, is proportional to the area of the heating surfaces and to the temperature difference between the two sides of the wall.

Furthermore, the capacity is influenced by the resistance to heat transfer largely caused by the existence of films and coatings on the heating surfaces. An important way of reducing the tendency to scaling, caused by coagulation of protein on the hot walls, is to use moderate steam temperatures, especially in the early stages of heating. Another measure is, of course, to introduce and enforce good routines for effective cleaning at regular intervals.

An entirely different type of heating device is the so-called Contherm apparatus, recently tried out in connection with fishmeal and oil manufacture. The results, so far at low rates of throughput, have been quite promising. The apparatus shown in Figure 7 consists of a vertical cylindrical heat exchanger provided with an agitator keeping the material in rapid movement, thus contributing to effective heat transfer. During rotation, the agitator blades (knives) are pressed against the surrounding heating surface in order to prevent the formation of scale. To reduce the viscosity of the material and to increase the rate of heat transfer, some stickwater should be added to the fish. Advantages of the Contherm heater are rapid heating with a holding time less than 2 min, effective temperature control, and quick and easy routines for dismantling and cleaning.

Another innovation is the tubular heater (pre-cooker) primarily designed and used for the utilization of waste heat, either from the evaporators or from the dryers. Because of the relatively low temperatures of these vapours or gases, they are particularly useful for preheating the raw material.

The heater consists of a set of tubes coupled together and surrounded by a cylindrical jacket. The raw material is moved through the tubes by pumping, and on its way it is heated by the hot gases circulating around the tubes, in the space between the latter and the jacket. In practical use, the raw material may reach temperatures in the range of 50° to 60 °C before it leaves the pre-heater and enters the conventional indirect steam cooker. Besides representing an important saving of energy this way of operation reduces the problem of scaling on the heating surfaces considerably, implying that the capacity of the heating units may be maintained at a high level for longer periods.

Pre-straining

One result of the heating process is that the oil and a major part of the water is released and to a large extent may be removed from the solids by simple draining. Removal of more liquid is achieved by subsequent treatment of the solid part in presses or centrifuges, or in a combination of the two.

To facilitate the functioning of the press, the liquid liberated in the cooker is drained from the coagulated fish pulp in a strainer conveyor or in a vibrating or rotary strainer. It is shown a strainer conveyor set at an incline between the cooker and the press.

It is designed on the same principle as that of a screw conveyor except that the lower end (that closer to the cooker) is fitted with an easily replaceable strainer in the shape of a half cylinder. Strainers with different sizes of perforations may be required for various types of fish. Figure 9 shows a vibrating strainer. The fundamental principle here is that the cooked material is conveyed to a strainer which is kept vibrating by an electric motor. The liquid phase passes through the strainer holes whereas the solid phase is vibrated along the surface of the strainer to an outlet.

Pressing

To ensure free drainage of liquid in the press, the material should be porous; that is, there should be many open channels in its mass for the passage of liquid. Cooked material from small and autolyzed fish will, as a rule, contain large quantities of fine particles (sludge) that tend to clog up these channels. In such cases, the porosity of the presscake may be improved by increasing the diameter of the holes of the pre-strainer. A greater part of the fines will then follow the liquid phase and not hamper the function of the press. To take advantage of this measure, the capacity of the decanters (desludging centrifuges) should be sufficient to handle the increased volume of sludge in the liquid.

The purpose of the press is to squeeze out as much liquid as possible from the solid phase. This is important not only to improve the oil yield and the quality of the meal, but also to reduce the moisture content of the presscake as far as possible, thereby reducing the fuel consumption of the dryers and increasing their capacity.

Two types of continuous press are used in the fishmeal industry; these are provided with either one or two screws. Both work on the principle of helical screw conveyors rotating in a tightly fitting cage, which is provided with perforations for the drainage of press liquid. The screws are made with a taper, thus ensuring that the volume between the flights is gradually reduced. This means that the material, during passage along the press, is subjected to increasing pressures and, as a consequence, additional amounts of liquid are expressed. The performance of the press is largely determined by the profile and the compression ratio of the screws, that is, the ratio between the flight volumes of the inlet and outlet flights. Whether standard screws based on fish of average nature and quality, or screws with a special profile and compression ration should be used, is a question for careful consideration and discussion with the press manufacturer.

Occasionally difficulties are experienced, particularly when processing soft and autolyzed fish. The press "slips", meaning that the screws rotate in the material without conveying it forward. This problem may be minimized by incorporating special devices in the single screw press; but the most efficient measure is to use two screws mounted side by side and rotating them in opposite directions.

For this reason the twin screw press has become the most commonly used type of press. Modern presses are very efficient dewatering devices, yielding presscakes with moisture contents as low as 50%.

The performance of the press may be regulated in two ways: one may adjust the level of cooked material in the hopper above the press, a high level resulting in higher pressure and consequently a more complete filling of the inlet screw flights; the other factor is the rate of revolution of the screws; increased speed means greater throughput and a shorter pressing time. How to adjust these two factors to obtain optimum performance is largely a matter of experience and skill.

Good performance of the press depends upon relatively tight fitting of the screw flights to the surrounding strainer plates. If the distance between the flight tops and the screens becomes too wide, for instance after long wear and tear, both the efficiency and the capacity will suffer; rebuilding and readjustment of the screw flights are then necessary. Another factor that needs continuous surveillance is the performance of the strainer plates. Regular inspection and cleaning is necessary to ensure that the holes are open and allow free escape of liquid. As pointed out earlier. temperature is a factor of great importance for the whole cooking and pressing operation. Basic information today indicates that moderate temperatures are preferable from the standpoint of release of oil and denaturation of protein. On the other side, high temperatures reduce the viscosity of the oil and tend to facilitate the flow from the solid phase. With the equipment we just have described, we must again rely on experimental data to establish optimum conditions for a particular raw material.

Processing problems may be encountered under two entirely different conditions. One relates to completely fresh fish that tends to retain more oil and water than desirable. For the time being, there is no solution to this problem except by resorting to one of the two equally deplorable measures; either by reducing the speed of the press and thereby the capacity of the whole plant. Or by storing the fish for a day or two before processing. thus leading to deterioration of quality.

The other situation occurs with soft and autolyzed fish. As mentioned in the introduction to this section, the answer to this problem is to bleed off in the pre-strainer more liquid and fines to be handled by the decanters. Some processors will often resort to the use of coagulating agents like formaldehyde, which help to solidify the material and improve the performance of the press. This, practice, however, should be restricted as far as possible because formaldehyde reacts with the essential amino acid lysine, and thereby reduces the nutritional quality of the protein. Calcium chloride ($CaCl_2$) has also been used as a hardener, but this practice was abandoned because it raised the chloride content of the meal to unacceptable levels, particularly in cases where stickwater is incorporated and whole meal produced.

Centrifugation instead of pressing

To separate solids from liquid by centrifugation is a standard operation in many industries including the fishmeal and oil industry. With the development of centrifuges that can handle materials with high contents of solids and at high rates of throughput, it 15 now possible to use decanters instead of presses to separate the solids from the liquid in cooked fish. (For a detailed description of the equipment see Figure 11 and Section 3.1.5). The advantages are several. First, it presents a simplification of the process. Secondly, centrifugation is a better known and more controllable unit operation than pressing and filtration. Thirdly, centrifugation is a much quicker process than pressing and significantly reduces the heat load on the material, a factor of importance for the manufacture of special products. Perhaps the most important advantage is the ability of the centrifuge to process soft and very fluid material where the press would fail completely. Better hygiene and simpler procedures for washing operations are further features on the plus side.

On the negative side one should note that the centrifuge will discharge the solids with a higher moisture content than the press. This means increased fuel consumption for the drying operation. Furthermore, the centrifuge tends to produce more emulsions and fines, causing problems in the subsequent separation of oil. water and sludge in the liquid phase.

Plants with decanters instead of presses are in practical operation in many parts of the world, generally with small or medium rates of throughput, ranging from 12 to 300 t of raw material per 24 h.

Although the use of decanters for the separation of solids and liquid in cooked fish material for the time being appears relatively unimportant, centrifugation is an interesting area where we may expect new developments. Combinations of press, strainer and centrifuge in various ways also open interesting possibilities which should prove worthwhile investigating.

Separation of press liquor

The liquor coming from the press and the pre-strainer consists of water and varying amounts of oil and dry matter. The oil content is related to the proportion of oil in the fish. The content of dry matter, occurring both in dissolved and suspended (finely dispersed) forms, varies with the size and quality of the fish and with the extent of mechanical handling prior to processing.

The quantity of press liquor will also vary with the nature and quality of the raw material, and increases particularly with advancing autolysis of the fish. Under average conditions one may estimate the volume of press liquor at about 70% of the raw material while the remaining 30% makes up the presscake.

The separation of the three fractions of the press liquor, sludge, oil and water, is based on their different specific gravities. If press liquor is left for some time in a tank, it will settle out in three layers: sludge at the bottom, water in between and oil at the top. In the early days of fish oil production, this method of settling under the influence of gravity alone was standard procedure. It had many drawbacks such as poor yield, impure fractions and, above all, it was extremely slow. With centrifugation we get several thousand times greater forces at our disposal, and the separation process may now be accomplished in seconds when compared with the hours required for the settling method.

An important prerequisite for efficient separation is high temperature, implying that the press liquor should be reheated to 90°-95°C before entering the centrifuges. This applies to sludge removal as well as to separation of oil and water.

The suspended solids are first to be removed. This is done in a horizontal centrifuge, a so- called decanter or desludger, the principle of which is shown in Figure 11. It consists of a partly cylindrical and partly conical rotor drum (bowl) and, inside this, a screw conveyor of the same shape. The press liquor is fed into the rotor where, by centrifugal force, it is thrown toward the bowl's periphery. The denser solids are rapidly precipitated along the inside rotor surface. The screw conveyor rotates with the bowl, but at a rate some 30 to 50 rpm faster than the speed of rotation of the drum; the deposited solids are thus scraped off continuously. Before being discharged, they are lifted out of the liquid phase and pass through a drying or dewatering zone.

The performance of the decanter may be controlled in two ways. It is possible to adjust the thickness of the liquid layer (a thick layer represents a longer zone and allows more time for clarification of the liquid) and, associated with this, there will be a correspondingly shorter zone of. sludge and less time for dewatering the solids. The reverse will, of course, be the case with a thin liquid layer. The other regulating parameter is the speed of the screw conveyor relative to that of the bowl. The higher the content of solids in the liquid the faster the conveyor should rotate in relation to the bowl in order to remove the precipitate. In addition to these parameters one may naturally influence performance by regulating the feed. Optimum conditions are dependent both on quantity and nature, specially particle size, of the solids in the liquid. Decanters are available in various sizes.

For smaller plants, the investment in a decanter may not be economically justified. In such cases a vibrating strainer, although less efficient, may be a cheaper but entirely satisfactory solution.

Separation of stickwater from oil takes place in vertical disc centrifuges, either of the nozzle type, which discharge the stickwater and remaining sludge continuously, or of the self cleaning type, which is often preferred. In the latter, the stickwater is continuously discharged, whereas the sludge is collected in the bowl and periodically ejected according to a timed programme which depends on the quantity and the nature of the sludge. The stickwater with a dry matter content of 6-9% is concentrated in evaporators. The sludge in most cases can be pumped to the presscake.

Figure 12 illustrates the principle of the self-cleaning disc centrifuge. The main component of the bowl is a stack of conical discs lying on top of each other at distances of 0.5 to 2 mm apart. The discs have a number of distribution holes to provide passages for the liquid from the bottom of the disc stack. The decanter liquid is fed from a control tube (l). The oil moves along the discs toward the centre and discharges through the holes in the nut (3). The stickwater moves toward the periphery and discharges behind the separating plate through the regulation ring (4). This is inter-changeable to adjust the separation. The sludge separates along the bowl periphery and is discharged through the bowl slot into the frame chute at regular intervals (2). Centrifuges are available with rates of throughput ranging from 500 to 25 000 litres/h.

Oil polishing

Oil polishing, carried out in special separators, is the final refining step done at the factory before the oil is pumped into storage. Polishing is facilitated by using hot water, which extracts impurities from the oil and thus ensures stability during storage.

The efficiency of separation depends upon both design and mode of operation of the centrifuges. The speed of separation depends upon the motility of the particles and upon the centrifugal force of the separator. Motility depends upon material properties, such as viscosity and specific gravity, which in turn depend upon temperature. Accordingly, good temperature control is required; the temperature of the feed should be maintained at about 95°C, but not less than 90°C. The centrifugal force is proportional to the angular velocity squared and to the radius of the centrifuge bowl, while the stress on the material of construction is proportional to the angular velocity squared and to the square of the radius. Centrifuges are designed to operate at high speeds and are, therefore, generally constructed with small radii. Centrifuges operating at about 5 000 rpm, yielding a centrifugal force of 5 000 x g (natural gravity), are generally used in the fishmeal industry.

Evaporation of stickwater

When decanters and separators have removed the major part of oil and suspended solids from the press liquid, we are left with the so-called stickwater. For all practical purposes, one may estimate the amount of stickwater at about 65% of the raw material. Besides water, stickwater will contain the following components:

- dissolved protein
- undissolved (suspended) protein
- residual oil
- minerals
- vitamins
- amines/ammonia

The content of residual oil will depend upon the efficiency of the separating process and should be as low as possible, certainly well below 1%. The other components, usually put together and called stickwater dry matter, amount to 5-6% for fresh fish, and correspond to nearly 20% of the yield of meal. After transportation and storage, specially of small fish caught during the feeding season and at high temperatures, the percentage of stickwater solids may rise to even higher values.

To recover the stickwater solids, one has to remove large quantities of water by evaporation and subsequent drying. This requires heat, and the question of heat economy and fuel consumption becomes, therefore, of paramount importance. The usage of heat may be influenced in various ways.

At the planning stage, one should carefully consider what type of evaporating plant to select, and particularly evaluate to what extent it is economically feasible to make use of waste heat, for instance that represented by the vapors in the exhaust gases from the dryer. Furthermore, there is the question of the number of evaporation stages, increasing numbers resulting in significant reductions in steam requirement. Typical figures for steam consumption are 0.6 to 0.65 kg, 0.4 to 0.45 kg and 0.2 to 0.35 kg steam per kilogramme water evaporated in double, triple and quadruple effect evaporators, respectively. On the other hand, the costs of construction rise with increase in the number of evaporator units. Generally, large catches of fish and long periods of operation speak in favour of cutting variable costs by investing in a greater number of evaporation stages. Other factors to be considered are, of course, local prices of fuel, electric energy and capital. For general guidance, it may be mentioned that double effects are used for rates of throughput of 30 to 150 t, triple ones for 200 to 400 t and quadruple for 500 t and more of raw fish per 24 h. If waste heat evaporators can be integrated into the system from the beginning, the picture will be changed significantly.

Selection of the operating conditions will also greatly affect the heat economy of the whole factory. Because multiple effect evaporation is a more economical way of removing water than one stage drying, it is specially important to achieve a high concentration of dry matter in the final stickwater concentrate before it is discharged from the evaporating plant and conveyed to the dryers for drying together with the presscake. The factor which largely determines how far you may concentrate the stickwater without running into trouble is the viscosity which rises steeply during the last stage of concentration. As increasing temperatures tend to make the concentrated stickwater less viscous, one may take advantage of this fact by completing the evaporation in the unit with the highest temperature. Another factor that can contribute greatly to the viscosity of the concentrate is the content of suspended solids (sludge), and great care should therefore be taken to keep this as low as possible, both by preventive measures and by efficient removal in decanters or sieves.

Although evaporation at high temperatures offers certain advantages, there are quality considerations that pull in the opposite direction. Some of the vitamins and amino acids are particularly sensitive to heat, and heating for any length of time above 130 °C should be discouraged, as this may lead to considerable losses of vitamin $B_{12}$and of the important amino acids, cystine, lysine and tryptophan. There are also other changes which take place under the influence of heat, such as degradation of the protein and evaporation of volatile components. Furthermore, there is the problem of discoloration. The significance of these changes should, of course, be carefully considered with special view to the marketing and end use of the products, before making the final decisions on equipment and operating conditions. Quality criteria will naturally centre around the nutritional value of the products but, with a view to a possible development toward products for human consumption, greater weight should be placed on sensory or organoleptic properties, like flavour and odour.

The conventional mode of operation is to feed dilute solution to the first stage operated above atmospheric pressure and to withdraw concentrate from the last, which is operated under reduced pressure. Some factories, however, prefer to feed dilute solution to the second stage and withdraw concentrate from the first. This subjects the concentrate to the highest available temperature in the system (generally above 100 °C), which has the following advantages: (1) reduction of viscosity of concentrate, (2) effective decomposition and removal of nitrogenous contaminants, for example nitrites and dimethyl nitrosamine, and oxides of sulphur if present, and (3) destruction of any pathogenic bacteria, including Salmonella, that might have infected the material in earlier stages. In some factories these effects are attained by treating the concentrated stickwater in separate heated pressure tanks.

During evaporation, solids are deposited on the hot surfaces impeding heat transfer and blocking the tubes. These deposits must be removed regularly, generally during shut-down periods (for example at weekends). Such removal necessitates chemical and mechanical treatments. Chemical treatment should be carried out when required, usually once a week. Mild steel evaporators may be cleaned with about 14% caustic soda solutions, recirculated at about 80°C for 5 h and left standing overnight. The apparatus is then emptied and thoroughly rinsed with water before the factory resumes normal operation. Mechanical scale removal in mild steel evaporators may be necessary several times a year. The weekly cleaning of stainless steel evaporators is usually done with stronger cleansing agents. Two hours' treatment with caustic soda will dissolve protein deposits, and a subsequent one hour's treatment with 5% nitric acid at 60°C will remove more firmly bound material. Mechanical cleaning of stainless steel evaporators should be avoided if possible but, if it should prove necessary, it should be carried out with great care to avoid scratches or other damage to the surface, which would result in loss of an important property of stainless steel, that is its smoothness. The smoother the surface the less firmly does the scale adhere to the surface. Monthly cleaning (caustic soda) on the steam side of the evaporators is also advisable. To prevent corrosion and facilitate cleaning, stainless steel tubes are recommended.

Oil separation from partly concentrated stickwater is practised by some manufacturers. The density of stickwater is higher in the concentrated than in the diluted state. This greater difference between the density of oil and concentrated stickwater produces an increase in the centrifugal potential and thus facilitates extra oil removal. Consequently, oil separation from the concentrate may lead to slightly leaner "whole" fish meal and will also increase the yield of oil. Oil separation seems to be more efficient after the second evaporation step than after later effects because of the lower viscosity of the concentrate at this stage. The separated oil tends to be rather dark and of less value than oil separated before concentration. High contents of sulphur are particularly harmful because some sulphur-containing compounds act as catalyst poisons during the hydrogenation process of the edible fats industry. "Concentrate oil" should, therefore, preferably be stored and sold separately.

Stickwater concentrate may be sold separately under the name "fish solubles". Because of the character of the product, however, the market is limited and is usually located not too far from the place of production. The most common way of utilizing stickwater concentrate is to mix it with the presscake and to dry the mixture to so-called "whole meal". The dry matter content of the concentrate is determined by the viscosity/dry matter relationship for the actual raw material, and may vary between 30% and 50%.

The commonly used evaporators operate with vertical tubes partly filled with boiling liquid, gradually giving off water vapour while moving upward in the tubes and into the chamber, separating liquid and vapours. The volume of liquid is quite large and it takes, therefore, a long time before the desired concentration has been reached, and concentrate starts flowing to the dryers. This time lag between pressing and concentrating is particularly undesirable and causes problems in factories with the intermittent supplies of raw material. In such cases, the falling film evaporator becomes of special interest because it operates with a short holding time and small liquid volume. Here the stickwater enters at the top of the tubes, which are heated from the outside by steam or hot gases. On its way down, water evaporates and the liquid becomes more and more concentrated, and finally ends up at the bottom as concentrate. The falling film evaporator also offers advantages in connection with energy saving systems using vapour recompression. So far, it has found limited use in the fish-meal industry largely because it requires special skill and attention from the operator. Because of its many advantages, however, the falling film evaporator is definitely a new alternative to consider for furture plants.

Drying

The purpose of the drying process is to convert the wet and unstable mixture of presscake, decanter sludge and concentration into a dry and stable fish meal. In practice, this means drying to a moisture content below 12%, which generally may be considered low enough to check microbial activity. This drying is done by heating the material to a temperature where the rate of evaporation of the water is considered satisfactory. Increasing the temperature will speed up the drying process. There are, however, certain critical limits t observe in order to avoid reduction of quality, especially of the protein.

With the equipment and conditions normally used in the fishmeal industry, the temperature of the drying material should not exceed 90°C in order not to impair nutrituional value.

A prerequisite for optimum drying conditions is that the material should be divided into relatively small pieces or particles so as to facilitate the escape of water vapour from below the surface into the air. The presscake is, therefore, passed through a wet mill where it is disintegrated by fast moving hammer heads mounted on a rotor. The presscake is beaten against a screen in the bottom provided with sharp edged, square holes. Figure 14 shows an example of a wet mill.

To avoid sticking and lump formation during one stage drying of presscake to which stickwater concentrate is added, thorough mixing is necessary. This may be achieved by heating the concentrate to a high temperature (about 100°C) before mixing. Such heating also serves to destroy bacteria, including Salmonella, if present. Mixing is generally performed in the screw conveyors, the concentrate preferably being added before disintegration of the presscake in the wet mill. In two stage drying the concentrate is added to the presscake between the dryers.

The type of dryer to be chosen will depend on several factors, i.e.:

i. the nature of the material to be dried,

ii. fuel economy,

iii. plant layout,

iv. plant capacity,

v. odour considerations.

The two main principles of drying are direct heat drying and indirect steam drying.

The direct rotary dryer, also called "flame dryer" or "direct hot air dryer", is used in the production of some 75% of fish meal in world trade. Heat for vaporization is provided by a current of flue gases diluted with secondary air, in direct contact with the fish material being dried. The direct action of the drying medium is both the strength and the weakness of the system. It represents the most efficient mode of heat and mass transfer but, unless operated properly, it may constitute a source of contamination of the product. Flue gas contaminants may range from products of incomplete combustion, for example particles of soot, to oxides of sulphur and nitrogen which are capable of reacting with nutritional components (such as protein and oil) in the meal. Manufacturers should therefore avoid the use of the crudest types of fuel, which may contain up to 1% by weight, or more, of sulphur and nitrogen. The temperature of combustion must be high enough to ensure complete fuel combustion; inlet air temperatures of 500° to 600°C are considered safe levels.

The direct rotary dryer is operated "in parallel" with respect to flow of air and fish material. Consequently, in addition to being the supplier of heat and the carrier of water vapour, the movement of the air also contributes to the transport of meal through the dryer. Accordingly, the air velocity must be adjusted so that the fish material is given sufficient dwell time in the dryer for proper drying. The drying time required, dependent upon process parameters and type of raw material, is established by careful adjustment for air flow coupled with measurement of fishmeal moisture during the start-up period.

The average dwell time in the dryer ranges between 10 and 20 min. By adjustment of the heat input, the evaporative capacity of the dryer can be varied. This is of importance to fishmeal manufacturers who have to face large fluctuations in quantity and quality of the raw material. Air temperature is a most important control parameter, since it affects the rate of evaporation. The temperature of the inlet air may range within wide limits without apparent damage to the meal. The reason for this is that the fish itself gets cooled through evaporation and normally does not exceed about 80°C, even though the air temperature may be several hundred degrees.

Adding stickwater concentrate to ground presscake, now almost universally practised, requires high concentration and good mixing to avoid problems of wet material sticking to metal surfaces; but these may be overcome by returning an adequate portion of dried material to the dryer inlet. Two stage drying, that is two rotary dryers operated in series, is practised in many instances on grounds of fuel economy and easier control of moisture content, and in these instances, the solubles are added to the partially dried presscake after the first drying stage.

Estimates of moisture content during drying are carried out manually, primarily and very roughly through "feel" of the fish meal to the hand, and by analysis. Fluctuations in moisture content of the end product are counteracted by manipulation of the hot air. Some factories employ indirect automatic moisture control based on continuous temperature measurement in the exit air. The temperature at this point is sensitive to variables in the process parameters of the system (for example flow, moisture content and homogeneity of the feed). Thus, when the exit air temperature drifts away from a fixed value an automatic device adjusts the oil burner and restores the desired condition. This control prevents gross variation in product quality.

A direct fired rotary dryer is illustrated in Figure 15. Rotating at a peripheral speed of about 1 m/s, the cylindrical compartment is equipped with horizontal and helical flights which provide cascading and good agitation of fish material, a large area of contact between fish and air and, accordingly, means for efficient dehydration. As stated, hot air is produced by mixing flue gases from oil gas or coal combustion with a stream of secondary air.

Because large seasonal catches of fish must be processed within relatively short periods of time, there has been a demand for large size rotary dryers in the industry capable of processing up to 1 000 t of raw material per day.

The indirect steam dryer works on the following principle: the mixture of presscake and stickwater concentrate is fed continuously into one end of the rotary apparatus, and is dried in direct contact with steam heated elements (tubes, discs, coils, etc.), emerging at the other end. A counter-current steam of air is blown through the dryer to facilitate removal of water vapour. The heat is transferred from the steam to the pulp through the heating surface, and rotary agitation of the pulp promotes the heat transfer.

The steam temperature (pressure) is limited by the dimensions and strength of materials of construction (discs, tubes or coils). A maximum steam temperature of 170°C corresponding to 6 atmospheres gauge pressure is most frequently used in steam dryers. Overall heat transfer is slower than in the direct dryer and a drying period of 30 min or longer is required. Figure 16 shows a rotary disc dryer. It consists of a steam heated stationary cylindrical jacket and a steam heated rotor equipped with steam heated double walled discs perpendicular to the rotor which provide good agitation and heat transfer to the fish meal. The evaporated water is removed by air drawn through the dryer by a centrifugal fan. There is an air dome on the top of the dryer to allow for the passage of air and water vapour. The entire drying process may be watched through inspection windows. The meal discharge is controlled by a gate valve at the discharge opening. These dryers normally use steam at a pressure of 6 kg/cm^2 and are designed to remove up to 2 700 kg water per hour, corresponding to about 300 t of raw fish per 24 h. One of the advantages of disc dryers is their ability to dry stickwater concentrate together with presscake without deposits forming on the heating surfaces.

Figure 17 illustrates the interior of a steam heated coil dryer. The main difference between this and the rotary disc dryer is the design of the rotating heating elements. The coils are mounted on a hollow shaft and supplied with steam at a pressure of about 7 kg/cm^2. These dryers are available with heating surface ranging between 20 and 400 m^2 and of handling materials corresponding to as much as 400 t of raw fish per 24 h.

The indirect tube dryer is a horizontal rotating cylinder with internal steam heated longitudinal tubes. To transfer heat from the tubes to the pulp and to remove evaporated water, air is drawn through the dryer.

The drying process in steam dryers is controlled by adjusting the steam pressure and by varying the rate of discharge of meal; the latter is done by adjusting the level of the material in the dryer.

In indirect dryers the fish meal is at no stage in contact with flue gases. The gentle transport of material through the stationary cylinder of a disc dryer appears to yield a product with a physical structure different from that of the rotary dryer. The particles are longer and less prone to spillage through grated cage floors and, consequently, the meal is more acceptable for mink feeding.

From a nutritional point of view, direct and indirect drying methods seem to yield products of equivalent value, The steam dryer has an important advantage in regard to air pollution. Owing to the small volume of effluent gases (30% or less of that of a direct heated rotary dryer of comparable capacity), odour abatement is much simpler. The great advantage of the direct dryer is the large drying capacity of each unit making for simpler machinery arrangement and cheaper installation for a given rate of throughput. With respect to heat economy, there is little to be said in favour of one or other type, when the efficiency of the steam dryer and that of the boiler are considered together.

Indirect hot air dryer have started to attract the interest of the fishmeal industry in recent years. They offer the same advantage as the steam dryers in the sense that the gases of combustion do not come into contact with the drying material. The hot air is generated in a heat exchanger where the gases from the combustion chamber move on one side of the heating surface and uncontaminated air on the other. After having transferred the heat to the current of clean air, the flue gases leave the system through the stack. With respect to the dryer itself, the performance is very much the same as that of the direct fired rotary dryer, the main difference being that clean, hot air replaces the current of flue gases with its various contaminants.

2. Milling and Storage of Fish Meal

Milling

Before milling, the meal should pass another vibrating sieve and magnet to remove extraneous matter, like pieces of wood, cloth, fish hooks, and nails, which might still be present. The purpose of milling is to facilitate uniform incorporation in feeds. A properly milled meal has an attractive appearance and is readily mixed into feed rations which require homogeneous blending.

Different users require fish meal of different particle sizes. The ideal in milling is to produce small particles averaging around 40 mesh Tyler screen and of as even a size as possible. In practice, however, there is a great variation in particle size, ranging from 10 mesh to over 100 mesh. Most purchasing specifications require the fish meal to pass through a 10 mesh screen, otherwise it is too coarse for uniform incorporation.

Production of excessive fines (particles below 150 mesh) should be avoided, for example by screening before milling and passing only the oversize particles through the mill. Large amounts of fines are undesirable for several reasons. They cause dusting when handled, sift through woven bags resulting in loss of weight and in pollution, cause compacting of bulk meal and tend to clog the nostrils of chickens eating the feed.

There are numerous types of dry mill on the market. In view of the need for high rates of throughput and ready access for cleaning, hammer mills have proved particularly suitable. The coarse meal is disintegrated by the impact of rapidly rotating hammers, pivoted on horizontal or vertical shafts. A grating is usually attached around the rotor which retains the meal until it is fine enough to pass through the perforations.

Figure 18 shows an example of a mill specially designed for grinding dried fish meal. The rotor consists of a central shaft to which a number of rotor plates are fastened perpendicular to the shaft. Between the rotor plates, hammers are held in position by bolts so that the hammers are removable. Special grinding plates are fixed inside the housing. At the lower part of the housing, there is a screen with round holes. By rotation of the rotor the material is hit by the hammers and the grinding plates and forced through the screens. The ground meal is cooled by air which is also used for transportation of the meal.

Addition of antioxidant

Reactive fish meals are "stabilized" by means of antioxidant immediately after manufacture, and may be stored in bulk or shipped as soon as they are cooled. The amount of antioxidant required for avoiding undue heating depends on the degree of reactivity of the oil (lipid unsaturation), and varies with fish species. Considerable excesses of ethoxyquin are, however, added for safety. Thus, South African pilchard meal (iodine value about 180) is "stabilized" with 400 ppm of ethoxyquin though 200 ppm would suffice, and herring meal (iodine value about 120) by 700 ppm of BHT or 200 ppm of ethoxyquin. The antioxidant is added immediately after drying. Anchoveta meal (iodine value about 190) is generally protected by 400 to 750 ppm of ethoxyquin.

Very careful control is necessary because of the small amount of antioxidant that is added to the fish meal and the need for even dispersion. For this reason the antioxidant is added to the meal in the screw conveyor leading from the dryer to the mill so that mixing can occur en route. Automatic controls are available for the addition of the antioxidant, complete with alarm bells and other devices to warn the factory personnel if anything is amiss, to avoid any fish meal being bagged without having been adequately treated. For ethoxyquin dosage it is essential to install proper automatic control and all fish meal passing the doser after the sounding of the alarm must be diverted. until the correct dosage has been re-established, and passed through the dosing system again.

In some factories the antioxidant is mixed with a constant amount of stickwater concentrate and this solution is then added to the presscake in the screw conveyor to the dryer. The effectiveness of the antioxidant is similar, whether added before or after drying. It must be stressed that stabilized meal retains a small trace of reactivity and is not completely stable. Nevertheless, the oil quality (energy value) is retained during prolonged storage; and, far more importantly, so is the protein quality of the meal, which otherwise could decrease through reaction with oxidized fish oil.

Weighing out fish meal into bags

Meal is frequently stored and transported in bags. The manner of weighing the finished meal varies and the degree of automation generally depends upon the processing capacity of the factory. At some fishmeal plants and particularly plants with a processing capacity of less than approximately 60 t of raw fish per 24 h, the bags are weighed and closed using a platform scale. More advanced and automatic methods are generally used in large plants. Stitchers for the bags range from portable sewing machines to large factory floor units. Self closing, valve bags for automatic bags for automatic filling and weighing machines are also used.

Bags: The bags usually hold 50 kg each, and may be open-ended and stitched, or with valves which are tucked in. Bag material ranges from hessian to multi-layer paper (with or without plastic lining), or sheet or woven plastic (low density polyethylene). The hessian bag, made from woven jute, is much used in tropical and sub-tropical countries. The relatively open fabric enables water vapour and heat to escape readily from the meal. The open fabric, however, has a number of disadvantages, such as:

the rapid entry of oxygen from the atmosphere facilitates a high rate of oxidation of residual oil in fish meal not treated with antioxidant. Indeed, the associated heat of reaction may be high enough to cause spontaneous combustion unless the meal is cured (that is allowed to oxidize slowly in single bags or rows of bags) or is protected by means of antioxidant;

- the meal is open to rodent and insect infestation and is readily contaminated through bacterial and fungal attacks;
- seepage of meal causes loss of product and pollution;
- under humid conditions the fish meal absorbs much water and may go mouldy and lumpy.

The paper bag (multi-layer, lined with polyethylene) is widely used by the industry. It keeps out insects and microbes and retards penetration of oxygen and water vapour from the atmosphere. As a result serious temperature increase may be avoided and there is negligible uptake of water vapour during storage.

The (solid sheet) plastic bag offers exceptionally good protection. A large variety of plastic materials (low density polyethylene, PVC, etc.) is now available, and bags may be tailored to suit specific requirements (for example, short- or long-term storage, rough handling, etc.). One of the most important criteria for the quality of plastic packing material is its resistance to penetration of water vapour and oxygen.

Oxidation and spontaneous heating are reduced to minute proportions only in tucked-in valve bags. If the bags are stitched the stitching holes allow sufficient entry of air to sustain some heating until the meal is cured, especially during handling of the bags. On the other hand, reactive meal in valved bags remains uncured and, if not treated with antioxidant, may be a serious source of spontaneous combustion if bulked immediately after removal from the bags.

Drop testing has indicated that polyethylene bags of 0.25 mm are stronger than multilayer paper bags. Nevertheless, transport and handling frequently result in more serious puncturing damage to polyethylene than to multi-layer paper bags. Also, the filled polyethylene bag is less rigid which, with its relatively low coefficient of friction, tends to make stowage more difficult and may more easily result in slipping and displacement during transport.

Palletizing: Wooden pallets holding about 1.5 t (30 bags) are often used to facilitate handling and stacking of bags after manufacture. Pallets may be stacked three high with fork lift trucks, after the meal has cooled to room temperature.

Cooling

Whether the freshly prepared fish meal is stabilized with ethoxyquin or not, it must be cooled to room temperature before it is stored in bulk. Bags of stabilized fish meal are left standing for a few days in single or double rows, or on singly-spaced pallets. Unbagged fish meal should be cooled by an air current or continually turned till cool. This can be done either in properly designed silos as, for example, in Norway and Denmark, or by turning the heaps of fish meal by means of bulldozers or conveyors as, for example, in the USA. Unstabilized reactive fish meal should be "cured" for 28 days, that is regularly turned if unbagged, or in small units if bagged, before stacking in bulk. For the first few days the bags should be placed on the floor apart from each other, or in single rows, depending on the initial reactivity of the meal; thereafter they may be stacked in double rows for the remaining 28-day period. Handling and storage on the floor involves a certain risk of Salmonellacontamination.

Pelletizing

Pelletized fish meal, mainly produced by large manufacturers, facilitates bulk storage and transport. The flow properties are improved and dusting is reduced. Pellets do not represent any essential space saving compared with bulk stored meal; nor are oxidation and spontaneous heating retarded by pelletizing. The bulk density of pellets is the same as that of fish meal (generally 600 to 700 kg/m^3). During handling, however, some of the pellets break and the broken pieces and meal formed in this way occupy the spaces between the pellets and thus increase the bulk density. Pellet diameters vary between 8 mm and 12 mm. Pelleting machines of 120 hp are capable of turning out about 5 t of pellets or more per hour.

Figure 19 illustrates the principle of the pellet press: (a) is the press mould (matrix), (b) the press rollers, (c) the distributor, (d) the knife, and (e) the pellets. The rollers press the meal through the holes of the matrix, and the knife cuts the pellets to the desired length. Usually, steam or hot stickwater concentrate is added to facilitate the formation of firm pellets. The pellets are usually made from hot meal emerging from the dryer and are cooled with fresh air in a cooling tower.

Storage

Storage methods for fish meal vary, depending upon many factors, including climatic conditions, production capacity, use of antioxidant and transport and marketing arrangements. Factories should have storage capacity for a reasonable buffer stock. In case of difficult marketing and shipping conditions, large storage ,capacities, for instance, sufficient to hold 30 days' production, may be required.

Fish meal must be protected from moisture and fishmeal stores must therefore be moisture proof. If necessary, the inner surface of the roof should be insulated or the stores provided with ceilings to avoid condensation and drip at night, with consequent localized mould growth and lumping in the fish meal. Only in arid regions may fish meal be stored out in the open. Fish meal must also be protected from undue self heating, whether it is treated with antioxidant or stored after curing. For this reason, bulk storage units should not exceed about 5 m in width.

Many different stacking arrangements for cooled bags are available from IAFMM. These arrangements are designed to limit the dimensions of the stacks, or to provide channels or "chimneys" through the stacks to carry away the small amount of residual heat generated. In this way, undue internal temperature rise is avoided and moisture migration from warm to cooler areas inhibited. Moisture migration can result in condensation, mould growth and lumping. Mould growth (as distinct from bacterial growth) may lead to spontaneous heating up to about 40°C at which temperature the moulds are destroyed; but at this elevated temperature spontaneous heating through oxidation of the oil may be accelerated, unless the fish meal is properly cured or stabilized with antioxidant.

Shipping stowage recommendations, also available from IAFMM, provide for athwartship twin columns of bags with 15 to 20 cm gaps between each tier, with cross dunnage.

Bulk storage: It is estimated that about half of the world's fish meal is stored in bulk in sheds and silos. Major manufacturers are taking an increasing interest in bulk storage because:

- all handling, from production to loading on transport vessels, becomes simpler, cheaper and results in a considerable saving in manpower used in maintenance, filling, weight control and stowage of bags;
- transport vessels and international receiving centres are being more and more geared to bulk handling, so that bulk storage in the factory is more compatible with modern overall handling systems.

In general, facilities for bulk storage are either of the open type (access of air through doors and other openings) or of the sealed type (space sealed off from the surroundings, as in special silos). The open type predominates in the fishmeal industry, and the ready availability of oxygen causes lipid oxidation of freshly made reactive meal not treated with antioxidant to proceed relatively rapidly. In order to dissipate the heat of reaction, the meal requires frequent aeration. This is particularly important during the initial storage stage when it is most reactive. A suitable practice of aeration involves conveying the meal from one storage space to another. If suitable antioxidant, however, is admixed with the meal, the latter is stabilized and considerably less aeration is required; indeed, only that which is necessary to cool the meal to room temperature. In open storage, special care should be given to pest and rodent control to prevent infection, for example by Salmonella.

The facilities in use for bulk storage of fish meal are the same as those employed in granaries, except for some modifications due to the "compacting" properties of the meal. The two most generally used systems are sheds, divided into several compartments, and silos.

Storage sheds may be of single or multistoried construction. Owing to the weight and the pressures exerted on the walls, such sheds are usually of single floor construction. Multistorage sheds should preferably be of concrete construction throughout, particularly the floors and outer walls.

Silos have found use in the fishmeal industry in recent years, not only because they offer good protection to meal during storage, but also because they increase flexibility in handling. This is perhaps the most important advantage of the silo over other types of storage facility.

Fish meal that is stored in specially designed silos may be kept in motion by continuously extracting it from the bottom and returning it to the top by means of automatic conveyor mechanisms. In this way, the meal is aerated for cooling and curing, and is also blended. It is also prevented from compacting and bridging. Generally, fish meal does not flow readily and tends to compact under pressure, especially at elevated temperature. The silos need special construction because of the relatively poor flow property of fish meal. Cylindrical silos are often preferred, but silos with square or rectangular cross-section are also satisfactory. The ease of discharge depends on the meal's physical properties and in particular on particle size distribution, especially fines content, moisture and fat contents, amount of solubles added, etc., and bridging is best prevented by assuring sufficient ventilation and circulation of meal during storage. The flow properties of fish meal can be improved by reducing the moisture content to 7% or by pelletizing.

The point of discharge at the bottom is the most critical part of the silo and is preferably constructed of steel. Depending on the cross-section of the silo, there may be several points of discharge.

The conical discharge section of cylindrical silos should have a slope at least 10° greater than the meal's angle of repose, which is of the order of 45°. Discharge chutes from corners of square silos are generally wedge-shaped, with three sides vertical and the fourth at an angle. A number of silos are normally arranged in one or more rows, which may take the form of a compact block. The multi-row arrangement is general for square silos since one silo will have walls in common with others, a saving in investment.

Blending silos constitute an integral part of the production and storage system. They are generally smaller than storage silos, and emphasis is placed on the efficiency and rate of their discharge and recirculation mechanisms to handle one to three days of peak fishmeal production. During blending, meals of various qualities are homogenized. This results in an overall quality increase because of:

- uniformity of the main constituents, protein, fat and water;
- more even grain size, which makes mixing of the meal into animal diets easier, and
- uniformity of colour, which improves the meal's market appeal.

Fish oil is stored in conventional tanks made from mild steel. The following design features are usually included:

- separation of residual water from the oil during storage necessitates water drainage at the tank bottom;
- in cold climates, the tanks must be equipped with steam-heated coils to keep the oil fluid during discharge;
- to facilitate good setting oil storage tanks are preferably insulated;
- to prevent contamination by sludge and water, unloading should be from a tube situated some way above the bottom of the tank.

3. Special Methods of Processing

In previous sections we have described the most common processing methods used in large and medium sized factories. For smaller factories and under special circumstances, simplified or entirely different installations may prove economical. In the following we shall give a few examples of different approaches, some of which aim at products with special properties.

Centrifugal methods

In these systems, the separation of the coagulated fish pulp is based on centrifugation instead of pressing. In one of these systems (Centrifish, DeLaval pat, Sweden), the fish is cooked in an indirect apparatus heated by flue gases, and the coagulated material separated into an oil-stickwater phase, containing some suspended sludge, and a solid phase in a decanter centrifuge. The liquid phase from the decanter is separated into oil, stickwater and sludge by means of self-cleaning centrifuges. The solids are dehydrated in an indirect tube dryer heated by flue gases.

The compact arrangement, which does not require a steam boiler, has made this system attractive for installation on board the ship. Without evaporation, however, about 20% to 30% of the solids is lost with the stickwater.

Packaged fishmeal plants

Packaged fishmeal plants of the traditional "press" type have also been developed for shipboard use. The processing capacities of compact plants range from 5 t to about 60 t of raw material per 24 h. Often the equipment is built in units to suit the requirements of the user and to facilitate installation. Such units may be cooker, press and dryer unit, separator and oil unit, evaporator unit, etc. Figure 20 illustrates a cooker, press and dryer unit, with the cooker at the top, the press below it and the dryer at the bottom. This unit has a processing capacity of 60 t of raw fish per 24 h.

Because of the relatively low installation costs and minimum operation attendance, packaged fishmeal plants are also of interest for stationary operation.

Meal production without cooking

A number of methods have been devised where the raw materials are dried without preliminary cooking. These are particularly suitable for lean fish (less than 3% fat) and offal, but may, in some circumstances, also be adapted to treat fatty fish species.

By the Schlotterhose method, lean fish is dried in two stages, using indirect steam dryers. In stage l, the moisture is generally reduced to about 55% under reduced pressure. The low pressure eases the removal of vaporized water and the relatively low temperature prevents meal from sticking and reduces lump formation. In stage 2, when sticking is less of a problem, drying is completed at atmospheric pressure.

Other methods (for example, Vega) employ two direct rotary dryers operated in series at atmospheric pressure. The feed to the first dryer may consist of a mixture of wet, raw, lean fish and semi-dried material. Due to decreased moisture content through mixing, this compounded feed is less prone to sticking than the wet raw material alone. The semi-dried material is dehydrated to the desirable moisture content in the second rotary dryer.

Dry rendering methods are applicable to some oily raw materials, mainly fish offal. If, however, the raw material contains highly reactive oils, for example from pilchard, there is a risk of inefficient oil removal during pressing.

In one such method (Hartman) the raw material is dehydrated batchwise in a single stage steam dryer to the desired moisture content, whereupon the dried material is treated in a hydraulic press. In order to achieve good pressing, the drying must be properly controlled. If the dried material contains less than 8% of water, there is a risk of inefficient oil removal. There is also a risk of oil darkening by direct contact between oil and meal during the drying process. Such degradation could limit the use of oil or require extra oil refinement and thus decrease profitability.

Simplified wet processing

A simplified version of the traditional wet processing method has been developed in recent years. It is applicable only to lean fish, where equipment for oil recovery serves no purpose. Furthermore, it is appropriate only for plants with low rates of throughput, where capital costs have to be reduced to the lowest possible level.

The process starts with cooking the raw material in a conventional indirect steam cooker. The cooked material is then dried to whole meal in a dryer with rotating heating surfaces. Finally, the meal is milled to the desired particle size, and bagged.

Since the evaporation of all the water removed is carried out in one stage, the heat economy is, of course, much inferior to that of the conventional method based on multiple stage evaporation of the stickwater. This drawback, however, should be more than compensated for by reduced investments in equipment.

Solvent extraction

For certain special purposes, fish meal from oily fish may not be used because of extremely low tolerance to fat of marine origin. In such cases, the fat content of the meal can be further reduced or practically eliminated by extraction with solvents. This treatment introduces an extra cost that has proved difficult to retrieve from the actual markets, represented particularly by the pig-feeding industry. Solvent extraction for the production of an odourless fish meal for human consumption has so far had little success, but new approaches in this field are still being tried.

Dry extraction of meal is carried out in reaction vessels where meal and heated solvent (for example, ethanol, isopropanol or a hydrocarbon) are given sufficient time for adequate oil extraction. Reduction of the oil content from 10% to 1% in menhaden, pilchard or anchovy meal would require about 4 litres of hexane per kilogramme of meal. More thorough extraction is achieved by successive re-extractions, up to five times or more, or by counter current extraction. After extraction, the meal is put through a process of solvent removal, which may be effected by steam stripping followed by hot air ventilation in a dehydrator. The permissible amount of residual solvent in the meal differs with the type of solvent used, the limits for chlorinated hydrocarbons being very low.

The liquid withdrawn from the extractor is fractionated to separate the fish oil, expel the water and recover the solvent. The oil is frequently dark and polymerized and not suitable for refining for human consumption. The recovered solvent requires additional treatment before it may be re-used for extraction. One of the best methods of removing undesirable fishy odours from the solvent is to pass the vapours through a special grade of active carbon. The active carbon can be regenerated by steaming when saturated with these unpleasant odours. Relatively high temperature boiling solvents, such as ethanol, isopropanol and certain hydrocarbons, are usually employed to obtain more thorough extraction of phospholipids and other odorous and flavorous substances, and on grounds of safety.

Wet extraction is the term given to the process of extracting oil and water from the wet raw material (fresh fish or presscake). From a technical standpoint, ethylene tetrachloride, which forms an azeotrope with water at 87.7°C, is suitable. At this temperature, water boils off together with the solvent, while the oil is simultaneously removed by extraction. As the tolerance for chlorinated solvents generally is very low, adequate removal of these solvents requires extra processing, details of which depend upon the ultimate use of the product.

Because of these objections to chlorinated solvents, alcohols like ethanol and isopropanol present more acceptable alternatives, particularly where products for human consumption are concerned.

Enzymatic treatment

Several methods have been developed for producing FPC or dried powder for animal feed by proteolysis of fresh whole fish. These methods are suitable for use on board the trawlers. In one method the fresh fish are pasteurized and put into a reaction vessel with papain (or other enzymes) at 55°C for about 1.5 h. The solubilized material is filtered and the cake dried and microground into a greyish product, which has a slight smell. The filtrate is treated in a separator and the oil withdrawn, while the aqueous phase (about 70%) is dehydrated in a spray dryer. The product is light yellow and has a faint smell.

Fish silage

There has been growing interest in producing liquid protein fodder from ensiled fish. Among the methods of ensilage used are treatments with mineral or organic acids (sulphuric or formic) and fermentation. In all methods the fish is ground and an example of production of acid silage is mixing with 3.5% of formic acid in a holding tank. The mass is hydrolysed to a stable product at a pH of 4. If mineral acids are used, the acidity must be reduced to pH 2 for stability. In micro-biological methods the ground fish is mixed with a source of carbohydrate (for example, 20% molasses) and a lactic acid producing micro-organism. Alcohol and lactic acid contribute to preservation. The silage products are generally dark brown semi-pastes, not unlike concentrated stickwater in appearance. They are convenient ingredients in liquid feeding compounds used locally, but the high water content makes long distance transport uneconomical.

Chapter 4

Cereal Processing

Food Extrusion

Food extrusion is a form of extrusion used in food processing. It is a process by which a set of mixed ingredients are forced through an opening in a perforated plate or die with a design specific to the food, and is then cut to a specified size by blades. The machine which forces the mix through the die is an extruder, and the mix is known as the extrudate. The extruder consists of a large, rotating screw tightly fitting within a stationary barrel, at the end of which is the die.

Extrusion enables mass production of food via a continuous, efficient system that ensures uniformity of the final product. Food products manufactured using extrusion usually have a high starch content. These include some pasta, breads (croutons, bread sticks, and flat breads), many breakfast cerealsand ready-to-eat snacks, confectionery, pre-made cookie dough, some baby foods, full-fat soy, textured vegetable protein, some beverages, and dry and semi-moist pet foods.

Process

In the extrusion process, raw materials are first ground to the correct particle size, usually the consistency of coarse flour. The dry mix is passed through a pre-conditioner, in which other ingredients are added depending on the target product; these may be liquid sugar, fats, dyes, meats or water.Steam is injected to start the cooking process, and the preconditioned mix (extrudate) is then passed through an extruder. The extruder consists of a large, rotating screw tightly fitting within a stationary barrel, at the end of which is the die.[1] The extruder's rotating screw forces the extrudate toward the die, through which it then passes. The amount of time the extrudate is in the extruder is the residence time.

The extruded product usually puffs and changes texture as it is extruded because of the reduction of forces and release of moisture and heat. The extent to which it does so is known as the expansion ratio. The extrudate is cut to the desired length by blades at the output of the extruder, which rotate about the die openings at a specific speed. The product is then cooled and dried, becoming rigid while maintaining porosity.

The cooking process takes place within the extruder where the product produces its own friction and heat due to the pressure generated (10–20 bar). The process can induce both protein denaturation and starch gelatinization, depending on inputs and parameters.

Many food extrusion processes involve a high temperature over a short time. Important factors of the extrusion process are the composition of the extrudate, screw length and rotating speed, barrel temperature and moisture, die shape, and rotating speed of the blades. These are controlled based on the desired product to ensure uniformity of the output.

Moisture is the most important of these factors, and affects the mix viscosity, acting to plasticize the extrudate. Increasing moisture will decrease viscosity, torque, and product temperature, and increase bulk density. This will also reduce the pressure at the die. Most extrusion processes for food processing maintain a moisture level below 40%, that is low to intermediate moisture.[2] High-moisture extrusion is known as wet extrusion, but it was not used much before the introduction of twin screw extruders (TSE), which have a more efficient conveying capability. The most important rheological factor in the wet extrusion of high-starch extrudate is temperature.

The amount of salt in the extrudate may determine the colour and texture of some extruded products. The expansion ratio and airiness of the product depend on the salt concentration in the extrudate, possibly as a result of a chemical reaction between the salt and the starches in the extrudate. Colour changes as a result of salt concentration may be caused by "the ability of salt to change the water activity of the extrudate and thus change the rate of browning reactions". Salt is also used to distribute minor ingredients, such as food colours and flavours, after extrusion; these are more evenly distributed over the product's surface after being mixed with salt.

The first extruder was designed to manufacture sausages in the 1870s. Packaged dry pasta and breakfast cereals have been produced via extrusion since the 1930s, and the method was applied to pet food production in the 1950s. It has also been incorporated into kitchen appliances, such as meat grinders, herb grinders, coffee grinders, and some types of pasta makers. A similar functional process occurs when using pastry bags.

Effects

Extrusion enables mass production of food via a continuous, efficient system that ensures uniformity of the final product. This is achieved by controlling various aspects of the extrusion process. It has also enabled the production of new processed food products and "revolutionized many conventional snack manufacturing processes".

The extrusion process results in "chemical reactions that occur within the extruder barrel and at the die".[6] Extrusion has the following effects:

- Inactivation of raw food enzymes
- Destruction of certain naturally occurring toxins
- Reduction of microorganisms in the final product
- Slight increase of iron-bioavailability
- Creation of insulin-desensitizing starches, which are a risk-factor for developing diabetes
- Loss of lysine, an essential amino acid necessary for developmental growth and nitrogen management
- Simplification of complex starches, increasing rates of tooth decay
- Increase of glycemic index of the processed food, as the "extrusion process significantly increased the availability of carbohydrates for digestion"
- Destruction of Vitamin A (beta-carotene)

Denaturation of proteins.

The material of which an extrusion die is made can affect the final product. Compared to stainless steel dies, a pasta machine with bronze dies produces a rougher surface. This is considered to give an improved taste, as it better retains pasta sauces. "Bronze die" pasta is labelled as such on retail packages, to indicate a premium product.

The effects of "extrusion cooking on nutritional quality are ambiguous", as extrusion may change carbohydrates, dietary fibre, the protein and amino acid profile, vitamins, and mineral content of the extrudate in a manner that is beneficial or harmful.

High-temperature extrusion for a short duration "minimizes losses in vitamins and amino acids". Extrusion enables mass production of some food, and will "denature antinutritional factors",such as destroying toxins or killing microorganisms. It may also improve "protein quality and digestibility", and affects the product's shape, texture, colour, and flavour.

It may also cause the fragmentation of proteins, starches, and non-starch polysaccharides to create "reactive molecules that may form new linkages not found in nature". This includes Maillard reactions which reduce the nutritional value of the proteins. Vitamins with heat lability may be destroyed. As of 1998, little is known about the stability or bioavailability of phytochemicalsinvolved in extrusion.[6] Nutritional quality has been found to improve with moderate conditions (short duration, high moisture, low temperature), whereas a negative effect on nutritional quality of the extrudate occurs with a high temperature (at least 200°C), low moisture (less than 15%), or improper components in the mix.

A 2012 research paper indicates that use of non-traditional cereal flours, such as amaranth, buckwheat or millet, may be used to reduce the glycemic index of breakfast cereals produced by extrusion. The extrudate using these cereal flours exhibits a higher bulk and product density, had a similar expansion ratio, and had "a significant reduction in readily digestible carbohydrates and slowly digestible carbohydrates". A 2008 paper states that replacing 5% to 15% of the wheat flour and white flour with dietary fibre in the extrudate breakfast cereal mix significantly reduces "the rate and extent of carbohydrate hydrolysis of the extruded products", which increased the level of slowly digested carbohydrates and reduced the level of quickly digested carbohydrates.

Products

Extrusion has enabled the production of new processed food products and "revolutionized many conventional snack manufacturing processes".

The various types of food products manufactured by extrusion typically have a high starch content. *Directly expanded* types include breakfast cereals and corn curls, and are made in high temperature, low moisture conditions under high shear. *Unexpanded* products include pasta, which is produced at intermediate moisture (about 40%) and low temperature. *Texturized* products include meat analogues, which are made using plant proteins ("textured vegetable protein") and a long die to"impart a fibrous, meat-like structure to the extrudate", and fish paste. Confectionery made via extrusion includes chewing gum, liquorice, and toffee.

Some processed cheeses and cheese analogues are also made by extrusion. Processed cheeses extruded with low moisture and temperature "might be better suited for manufacturing using extrusion technology" than those at high moisture or temperature. Lower moisture cheeses are firmer and chewier, and cheddar cheese with low moisture and an extrusion temperature of 80°C was preferred by subjects in a study to other extruded cheddar cheese produced under different conditions.[16] An extrudate mean residence time of about 100 seconds can produce "processed cheeses or cheese analogues of varying texture (spreadable to sliceable)". Other food products often produced by extrusion include some breads (croutons, bread sticks, and flat breads), various ready-to-eat snacks, pre-made cookie dough, some baby foods, somebeverages, and dry and semi-moist pet foods. Specific examples include macaroni, Fig Newtons, jelly beans, sevai, and some french fries. Extrusion is also used to modify starch and to pelletanimal feed.

How to Enhance the Nutritive Value of Foods

Good food with adequate nutrients is important for one's physical and mental health. Eating a variety of foods does not mean that the individual is assured of the entire essential nutrients especially vitamins and minerals needed for the body. There are lot of methods like combination, sprouting, fermentation, parching, puffing, parboiling, liming are used to enhance the nutritive value of the foods.

Combination

Combination of foods provides variety to the diet as well as increases the nutritive value of the foods. Pulses are major source of proteins for vegetarians but they lack in essential amino acid methionine but cereals provide this amino acid. Like wise cereals also provide proteins but they lack in amino acid lysine. This amino acid is present in pulses. If the two are combined the protein quality of the both is superior to either cereals or pulses taken alone. Some of the cereal pulse combinations are idli, dhokla, khichadi, mixing wheat flour with Bengal gram flour, soy flour and wheat flour, cereal, pulse, and vegetable. Pulao and aloo paratha are commonly used food preparation of this combination. Combination supplies balanced amounts of protein, carbohydrate, and minerals. Leafy vegetables are good source of vitamin A, calcium and iron for e.g. spinach paratha, fenugreek paratha, spinach dhal etc. Milk and wheat, milk and rice, milk and wheat germ also provides good supply of nutrients especially protein, calcium, vitamin E. Animal protein like meat, fish and egg should be taken along with vegetables to get the maximum of protein, minerals and vitamins.

Sprouting

In this method the raw grains or beans are soaked in water for 10 -12 hours with equal amount of water till almost all the water is absorbed and the seeds swell up to their maximum size. The soaked grains should be tied or covered in a wet muslin cloth and kept in warm place the cloth should be kept moist till the shoots appear. The sprouted grain may be eaten raw with salt, lemon juice, and other vegetables like onion, tomato, cucumber, cabbage, and carrot either raw or cooked. Wheat, green gram, cowpea, soybean, bengal gram and fenugreek seeds are usually sprouted. Germination or sprouting enhances the vitamin C content of the pulses and cereals. Niacin content is increased to 60 - 100%. Iron gets loosened from its bound form and becomes easily available. Digestibility of sprouted products is increased because the cell walls carrying carbohydrates and proteins are broken down during the process. Digestibility increases due to production of amylase. Sprouted foods can be converted to amylase rich food. Sprouting of fenugreek reduces its bitter taste.

Parboiling

This method is used mainly for the paddy. The paddy is soaked in water for some time and then it is steamed or boiled. It is milled after drying to remove the brown husks. In parboiling most of the nutrients in the husk are absorbed by the grains. Especially the water-soluble vitamins of the B group. Thiamine and riboflavin are doubled than the polished rice. While cooking parboiled rice, the loss of nutrients is much less than in case of polished rice.

Fermentation

Tiny micro-organisms that are naturally present in foods bring about fermentation. Making of curd from milk is a good example of fermentation. Milk sets into curd due to the presence of the lactobacillus. Lactose is converted into lactic acid, which is sour in taste. Fermented food is soft easy to digest and provides variety in texture. Fermentation increases thiamine, riboflavin, and niacin to double the original amounts. Nutrients such as vitamin C, methionine, and folic acid are also increased during this process. Iron gets released from its bound form during fermentation and becomes easily absorbable.

Parching

Parching method is used for wheat, maize, rice, bengal gram, and peas. These food items are made moist and suddenly heated. The water escapes by puffing out the grain. Popcorn, puffed rice or wheat are commonly used parched products. They make a good break fast cereals with sugar and milk. Even ladoos can be made with puffed cereal with jaggery. Starch becomes more digestible after parching. The cell walls of the starch grain rupture and the large chemical structure of the starch break down into smaller ones. Some loss of lysine in the protein does occur because of the high temperature necessary for parching.

Cooking

Cooking does lose some of the nutrients but some are enhanced after cooking. Biotin a vitamin of the B group is present in fair amounts in egg yolk but the white of the egg contains a substance called avidin, which prevents availability of biotin. Avidin is completely destroyed by cooking and biotin is made available to the body. Like wise during cooking trypsin inhibitors are destroyed rendering the proteins present in foods gets better digestibility. The wet methods of cooking lead to comparatively greater losses than the dry methods. The tamarind when added in cooking act as vitamin preservative.

Food-Based Approaches to Meeting Vitamin and Mineral Needs

Dietary patterns have varied over time depending on the agricultural practices and the climatic, ecologic, cultural, and socio-economic factors, which determine available foods. At present, virtually all dietary patterns adequately satisfy or even exceed the nutritional needs of population groups. This is true except where socio-economic conditions limit the capacity to produce and purchase food or aberrant cultural practices restrict the choice of foods. It is thought that if people have access to a sufficient quantity and variety of foods, they will meet their nutritional needs. The current practice of evaluating nutritive value of diets should include not only energy and protein adequacy but also the micronutrient density of the diet.

A healthy diet can be attained in more than one way because of the variety of foods, which can be combined. It is thus difficult to define the ranges of intake for a specific food, which should be included in a given combination to comply with nutritional adequacy. In practice, the set of food combinations which is compatible with nutritional adequacy is restricted by the level of food production sustainable in a given ecologic and population setting. In addition, there are economic constraints, which limit food supply at household level. The development of food-based dietary guidelines (FBDGs) by the FAO and WHO (*1*) recognises this and focuses on the combination of foods that can meet nutrient requirements rather than on how each specific nutrient is provided in adequate amounts.

The first step in the process of setting dietary guidelines is defining the significant diet-related public health problems in a community. Once these are defined, the adequacy of the diet is evaluated by comparing the information available on dietary intake with recommended nutrient intakes (RNIs). Nutrient intake goals under this situation are specific for a given ecologic setting, and their purpose is to promote overall health, control specific nutritional diseases (whether they are induced by an excess or deficiency of nutrient intake), and reduce the risk of diet-related multi-factorial diseases. Dietary guidelines represent the practical way to reach the nutritional goals for a given population. They take into account the customary dietary pattern and indicate what aspects should be modified. They consider the ecologic setting, socio-economic and cultural factors, and biologic and physical environment in which the population lives.

The alternative approach to defining nutritional adequacy of diets is based on the biochemical and physiologic basis of human nutritional requirements in health and disease. The quantitative definition of nutrient needs and its expression as RNIs have been important instruments of food and nutrition policy in many countries and have focused the attention of international bodies. This nutrient-based approach has served many purposes but has not always fostered the establishment of nutritional and dietary priorities consistent with the broad public health priorities at the national and international levels. It has permitted a more precise definition of requirements for essential nutrients when establishing RNIs but unfortunately has often been narrowly focused, concentrating on the precise nutrient requirement amount and not on solving the nutritional problems of the world. In contrast to RNIs, FBDGs are based on the fact that people eat food, not nutrients. As illustrated in this chapter, the notion of nutrient density is helpful for defining FBDGs and evaluating the adequacy of diets. In addition, they serve to educate the public through the mass media and provide a practical guide to selecting foods by defining dietary adequacy (1).

Advice for a healthy diet should provide both a quantitative and qualitative description of the diet for it to be understood by individuals, who should be given information on both size and number of servings per day. The quantitative aspects include the estimation of the amount of nutrients in foods and their bio-availability in the form they are actually consumed. Unfortunately, available food composition data for most foods currently consumed in the world are incomplete, outdated, or insufficient for evaluating true bio-availability. The qualitative aspects relate to the biologic utilisation of nutrients in the food as consumed by humans and explore the potential for interaction among nutrients. Such an interaction may enhance or inhibit the bio-availability of a nutrient from a given food source.

Including foods in the diet, which have high micronutrient density - such as pulses or legumes, vegetables (including green leafy vegetables), and fruits - is the preferred way of ensuring optimal nutrition including micronutrient adequacy for most population groups. Most population groups afflicted by micronutrient deficiency largely subsist on refined cereal grain or tuber-based diets, which provide energy and protein (with improper amino acid balance) but are insufficient in critical micronutrients. The recent interest in the role of phyto-chemicals and antioxidants on health and their presence in plant foods lend further support to the recommendation for increasing vegetables and fruit consumed in the diet. The need for dietary diversification is supported by the knowledge of the interrelationships of food components, which may enhance the nutritional value of foods and prevent undesirable imbalances, which may limit the utilisation of some nutrients. For example, fruits rich in ascorbic acid will enhance the absorption of ionic iron.

If energy intake is low (<8.368 MJ/day), for example, in the case of young children, sedentary women, or the elderly, the diet may not provide vitamin and mineral intakes sufficient to meet the RNIs. This situation may be of special relevance to the elderly, who are inactive, have decreased lean body mass, and typically decrease their energy intake. Young children, pregnant women, and lactating women, who have greater micronutrient needs relative to their energy needs, will also require increased micronutrient density.

The household is the basic unit for food consumption under most settings, and if there is sufficient food, individual members of the household can consume a diet with the recommended nutrient densities and meet their specific RNIs. However, appropriate food distribution within the family must be considered to ensure that children and women receive adequate food with high micronutrient density. Household food distribution must be considered when establishing general dietary guidelines and addressing the needs of vulnerable groups in the community. In addition, education detailing the appropriate storage and processing of foods to prevent micronutrient losses at the household level is important.

Dietary Diversification when Consuming Cereal and Tuber-based Diets (Rice, Corn, Wheat, Potato, and Cassava)

Dietary diversification is important to improve the intake of critical nutrients. The micronutrients selected discussed here, although limited in number, are of public health relevance or serve as markers for overall micronutrient intake. The chapters on individual nutrients will provide further details on food-related considerations for micronutrient adequacy. The nutrients selected for discussion below include some of the nutrients, which are most difficult to obtain in cereal and tuber-based diets. Nutrient deficiencies of vitamin A, iron, and zinc are widespread.

Vitamin A

The vitamin A content of most staple diets can be significantly improved with the addition of a relatively small portion of plant foods rich in carotenoids, the precursors of vitamin A. For example, a usual portion of cooked carrots (50 g) added to a daily diet, or 21 g of carrots per 4.184 MJ, provides 500 mg retinol equivalents, which is the recommended nutrient density for this vitamin. The biologic activity of pro-vitamin A varies among different plant sources, and fruits and vegetables such as carrots, mango, papaya, and melon contain large amounts of nutritionally active carotenoids, (2, 3). Green leafy vegetables such as ivy gourd have been successfully used in Thailand as a source of vitamin A, and carotenoid-rich red palm oil serves as an easily available and excellent source of vitamin A in other countries. Consequently, a regular portion of these foods included in an individual's diet may provide 100 percent or more of the daily requirement for retinol equivalents. Vitamin A is also present in animal food sources in a highly bio-available form. Therefore it is important to consider the possibility of meeting vitamin A needs by including animal foods in the diet. For example, providing minor amounts of fish or chicken liver (20-25 g) in the diet provides more than the recommended vitamin A nutrient density for virtually all age and sex groups.

Vitamin C

A real gain in vitamin C intake can be achieved by including citrus fruit or other foods rich in ascorbic acid in the diet. For example, an orange or a small amount of other vitamin C-rich fruit (60 g of edible portion) provides the recommended ascorbic acid density. Adding an orange to a potato-based diet increases the level of vitamin C threefold. Other good vitamin C food sources are guava, amla, kiwi, cranberries, strawberries, papaya, mango, melon, cantaloupe, spinach, Swiss chard, tomato, asparagus, and Brussels sprouts. All these foods, when added to a diet or meal in regular portion sizes, will significantly improve the vitamin C density. Because ascorbic acid is heat labile, minimal cooking (steaming or stir-frying) is recommended to maximise the bio-available nutrient. The significance of consuming vitamin C with meals will be discussed relative to iron absorption.

Folate

Folate is now considered significant not only for the prevention of macrocytic anaemia, but also for normal foetal development. Recently, this vitamin was implicated in the maintenance of cardiovascular health and cognitive function in the elderly. Staple diets consisting largely of cereal grains and tubers are very low in folate but can be improved by the addition of legumes or green leafy vegetables. For example, a regular portion of cooked lentils (95 g) added to a rice-based diet can provide an amount of folate sufficient to meet the desirable nutrient density for this vitamin. Other legumes such as beans and peas are also good sources of this vitamin, but larger portions are needed for folate sufficiency (100 g beans and 170 g peas). Cluster bean and colacasia leaves are excellent folate sources used in the Indian diet. Another good source of folate is chicken liver; only one portion (20-25 g) is sufficient to meet the desirable nutrient density for folate and vitamin A simultaneously. The best sources of folate are organ meats, green leafy vegetables, and sprouts. However, 50 percent or more of food folate is destroyed during cooking. Prolonged heating in large volumes of water should be avoided, and it is advisable to consume the water used in the cooking of vegetables.

Iron and zinc

Minerals such as iron and zinc are low in cereal and tuber-based diets, but the addition of legumes can slightly improve the iron content of those diets. However, the bio-availability of this non-heme iron source is low. Therefore, it is not possible to meet the recommended levels of iron and zinc in the staple-based diets through a food-based approach unless some meat, poultry, or fish is included. For example adding a small portion (50 g) of meat, poultry, or fish will increase the total iron content as well as the amount of bio-available iron. For zinc the presence of a small portion (50 g) of meat, poultry, or fish will secure dietary sufficiency of most staple diets.

The consumption of ascorbic acid along with the food rich in iron will enhance absorption. There is a critical balance between enhancers and inhibitors of iron absorption. Nutritional status can be improved significantly by educating households on food preparation practices, which minimise the consumption of inhibitors of iron absorption; for example, the fermentation of phytate-containing grains before the baking of breads to enhance iron absorption.

How to accomplish dietary diversity in practice

It is essential to work on strategies, which promote and facilitate dietary diversification to achieve complementarity of cereal or tuber-based diets with foods rich in micronutrients in populations with limited economics or limited access to food. A recent FAO and International Life Sciences Institute (*4*) publication proposed strategies to promote dietary diversification within the implementation of food-based approaches. These strategies, which follow, have been adapted or modified based on the discussions held in this consultation:

1. Community or home vegetable and fruit gardens.

These projects should lead to increased production and consumption of micronutrient-rich foods (legumes, green leafy vegetables, and fruits) at the household level. The success of such projects requires a good knowledge and understanding of local conditions as well as the involvement of women and the community in general. These are key elements for supporting, achieving, and sustaining beneficial nutritional change at the household level. Land availability and water supply may present common constraints, which require local government intervention or support before they are overcome. The educational effort should be directed towards securing appropriate within-family distribution, which considers the needs of the most vulnerable members of the family, especially infants and young children. Separate FBDGs for vulnerable groups, such as pregnant and lactating women, children, and the elderly, should be developed.

2. Production of fish, poultry, and small animals (rabbits, goats, and guinea pigs).

These are excellent sources of highly bio-available essential micronutrients such as vitamin A, iron, and zinc. The production of animal foods at the local level may permit communities to access foods which otherwise are not available because of their high costs. These types of projects also need some support from local governments or non-governmental organizations to overcome cost constraints of programme implementation, including the training of producers.

3. Implementation of large-scale commercial vegetable and fruit production.

The objective of this initiative is to provide micronutrient-rich foods at reasonable prices through effective and competitive markets, which lower consumer prices without reducing producer prices. This will serve predominantly the urban and non-food-producing rural areas.

4. Reduction of post-harvest losses of the nutritional value of micronutrient-rich foods, such as fruits and vegetables.

Improvement of storage and food-preservation facilities significantly reduces post-harvest losses. At the household level, the promotion of effective cooking methods and practical ways of preserving foods (solar drying of seasonal micronutrient-rich foods such as papaya, grapes, mangoes, peaches, tomatoes, and apricots) may significantly increase the access to bio-available micronutrient-rich foods. At the commercial level, grading, packing, transport, and marketing practices reduce losses, stimulate economic growth, and optimise income generation.

5. ***Improvement of micronutrient levels in soils and plants, which will improve the composition of plant foods and enhance yields.***

Current agricultural practices can improve the micronutrient content of foods through correcting soil quality and pH and increasing soil mineral content depleted by erosion and poor soil conservation. Long-term food-based solutions to micronutrient deficiencies will require improvement of agricultural practices, seed quality, and plant breeding (by means of a classical selection process or genetic modification).

The green revolution made important contributions to cereal supplies, and it is time to address the need for improvements in the production of legumes, vegetables, fruits, and other micronutrient-rich foods. FBDGs can serve to reemphasise the need for these crops.

It is well recognised that the strategies proposed to promote dietary diversity need a strong community-level commitment. For example, the increase in price of legumes associated with decreased production and lower demand needs to be corrected. The support of local authorities and government may facilitate the implementation of such projects because these actions require economic resources, which sometimes are beyond the reach of the most needy.

Practices which will Enhance the Success of Food-based Approaches

To achieve dietary adequacy of vitamin A, vitamin C, folate, iron, and zinc by using food-based approaches, food preparation and dietary practices must be considered. For example, it is important to recommend that vegetables rich in vitamin C, folate, and other water-soluble or heat-labile vitamins be minimally cooked in small amounts of water. For iron bio-availability it is essential to reduce the intake of inhibitors of iron absorption and to increase the intake of enhancers of absorption in a given meal. Following this strategy, it is recommended to increase the intake of: germinated seeds, fermented cereals, heat-processed cereals, meats, and fruits and vegetables rich in vitamin C and to encourage the consumption of tea, coffee, chocolate, or herbal teas at times other than with meals. Consumption of flesh foods improves zinc absorption whereas it is inhibited by consumption of diets high in phytate, such as diets based on unrefined cereal. Zinc availability can be estimated according to the phytate-to-zinc (molar) ratio of the meal.

This advice is particularly important for people who consume cereal and tuber-based diets. These foods constitute the main staples for most populations of the world, populations that are also most at risk for micronutrient deficiencies. Other alternatives - fortification and supplementation - have been proposed as stopgap measures when food-based approaches are not feasible or are still in progress.

There is a definite role for fortification in meeting iron, folate, iodine, and zinc needs. Fortification and supplementation should be seen as complementary to food-based strategies and not as a replacement. Combined, all these strategies can go a long way toward stabilising the micronutrient status of populations at risk. Food-based approaches usually take longer to implement but once established are truly sustainable.

Delineating the role of Supplementation and Food Fortification for Nutrients which cannot be Supplied by Regular Foods

Under ideal conditions of food access and availability, food diversity should satisfy micronutrient and energy needs of the general population. Unfortunately, for many people in the world, the access to a variety of micronutrient-rich foods is not possible. As demonstrated in our analysis of cereal and tuber-based diets, micronutrient-rich foods including small amount of flesh foods and a variety of plant foods (vegetables and fruits) are needed daily. This may not be realistic at present for many communities living under conditions of poverty. Food fortification and food supplementation are important alternatives that complement food-based approaches to satisfy the nutritional needs of people in developing and developed countries.

Fortification

Fortification refers to the addition of nutrients to a commonly eaten food (the vehicle). It is possible for a single nutrient or group of micronutrients (the fortificant) to be added to the vehicle, which has been identified through a process in which all stakeholders have participated. This strategy is accepted as sustainable under most conditions and often is cost effective on a large scale when successfully implemented. Iron fortification of wheat flour and iodine fortification of salt is examples of fortification strategies with excellent results (*6*).

There are at least three essential conditions that must be met in any fortification programme(*6, 7*): the fortificant should be effective, bio-available, acceptable, and affordable; the selected food vehicle should be easily accessible and a specified amount of it should be regularly consumed in the local diet; and detailed production instructions and monitoring procedures should be in place and enforced by law.

Iron fortification

Food fortification with iron is recommended when dietary iron is insufficient or the dietary iron is of poor bio-availability, which is the reality for most people in the developing world and for vulnerable population groups in the developed world. Moreover, the prevalence of iron deficiency and anaemia in vegetarians and in populations of the developing world which rely on cereal or tuber foods is significantly higher than in omnivore populations.

Iron is present in foods in two forms, as heme iron, which is derived from flesh foods (meats, poultry, and fish), and as non-heme iron, which is the inorganic form present in plant foods such as legumes, grains, nuts, and vegetables (*8, 9*). Heme iron is highly (20-30 percent) absorbed and its bio-availability is relatively unaffected by dietary factors. Non-heme iron has a lower rate of absorption

(2-10 percent), depending on the balance between iron absorption inhibitors (phytates, polyphenols, calcium, and phosphate) and iron absorption enhancers (ascorbic and citric acids, cysteine-containing peptides, ethanol, and fermentation products) present in the diet (*8, 9*). Because staple foods around the world provide predominantly non-heme iron sources of low bio-availability, the traditionally eaten staple foods represent an excellent vehicle for iron fortification. Examples of foods, which have been fortified, are wheat flour, corn (maize) flour, rice, salt, sugar, cookies, curry powder, fish sauce, and soy sauce (*8*). Nevertheless the beneficial effects of consumption of iron absorption enhancers have been extensively proven and should always be promoted (i.e., consumption of vitamin C-rich food together with the non-heme iron source).

Iodine fortification

Iodine is sparsely distributed in the Earth's surface and foods grown in soils with little or no iodine lack an adequate amount of this micronutrient. This situation had made iodine deficiency disorders exceedingly common in most of the world and highly prevalent in many countries before the introduction of salt iodisation (*10*). Only foods of marine origin are naturally rich sources of iodine. Salt is a common food used by most people worldwide, and the establishment of an well-implemented permanent salt-iodisation programme has been proven to eradicate iodine deficiency disorders. Universal salt iodisation is the best way to virtually eliminate iodine deficiency disorders by the year 2000 (*4*).

However, salt iodisation is not simply a matter of legislating mandatory iodisation of salt. It is important to determine the best fortification technique, co-ordinate the implementation at all salt production sites, establish effective monitoring and quality control programmes, and measure iodine fortification level periodically. The difficulties in implementing salt iodisation programmes arise primarily when the salt industry is widely dispersed among many small producers. The level of iodine fortification usually lies between 25 and 50 mg/kg salt. The actual amount should be specified according to the level of salt intake and magnitude of deficit at the country level, because iodine must be added within safe and effective ranges. Additionally, it is very important to implement a monitoring plan to control the amount of iodine in the salt at the consumer's table, (*10, 11*). United Nations agencies responsible for assisting governments in establishing iodisation programmes should provide technical support for programme implementation, monitoring, and evaluation to ensure sustainability.

Zinc fortification

The body depends on a regular zinc supply provided by the daily diet because stores are quite limited. Food diversity analysis demonstrates that it is virtually impossible to achieve zinc adequacy in the absence of a flesh food source. Among flesh foods, beef is the best source of zinc and is followed by poultry and then fish. Zinc fortification programmes are being studied, especially for populations, which consume predominately plant foods. Fortification of cereal staple foods is a potentially attractive intervention, which could benefit the whole population as well as target the vulnerable population groups of children and pregnant women. Such addition of zinc to the diet would perhaps decrease the prevalence of stunting in many developing countries with low-zinc diets, because linear growth is affected by zinc supply.

Folic acid fortification

The recommended nutrient density by the developers of the FAO/WHO (*1*) FBDGs for folic acid is 200 mg/4.184 MJ. Although this reference value is higher than other standards of reference, the increase in folic acid consumption by women of childbearing age is very important: it may improve birth weight and reduce the prevalence of neural tube defects by 50 percent. Elevated plasma homo-cysteine levels are considered to be an independent risk factor for heart disease; a higher intake of folic acid may also benefit the rest of the population because it may lower homo-cysteine levels in adults. In addition, folate may improve the mental condition of the elderly population.

Although the desirable folic acid density may be achieved through dietary diversity, it requires the daily presence of organ meats, green leafy vegetables, pulses, legumes, or nuts in the diet. Most population groups may not easily reach the appropriate level of folic acid consumption; therefore, folic acid fortification has been recommended. The United States initiated mandatory folic acid fortification of cereal-grain products in January 1998. The fortification level approved in the United States is 140 mg/100 g product, which will increase the average woman's intake by only 100 mg/day. This amount is considered safe (a dose, which will not mask pernicious anaemia, which results from vitamin B12 deficiency,) but it may be ineffective in lowering the occurrence of neural tube defects.

Supplementation

Supplementation refers to periodic administration of pharmacologic preparations of nutrients as capsules or tablets or by injection when substantial or immediate benefits are necessary for the group at risk. As established at the *International Conference on Nutrition*, nutritional supplementation should be restricted to vulnerable groups, which cannot meet their nutrient needs through food (women of childbearing age, infants and young children, elderly people, low socio-economic groups, displaced people, refugees, and populations experiencing other emergency situations). For example, iron supplementation is recognised as the only option to control or prevent iron deficiency anaemia in pregnant women. Supplementation with folic acid should be considered for women of childbearing age who have had a child with neural tube defect to prevent recurrence.

Food-based dietary guidelines

Food-based dietary guidelines (FBDGs) are an instrument of and expression of food and nutrition policy and should be based directly on diet and disease relationships of particular relevance to the individual country. Their primary purpose is to educate healthcare professionals and consumers about health promotion and disease prevention. In this way priorities in establishing dietary guidelines can address the relevant public health concerns whether they are related to dietary insufficiency or excess. In this context, meeting the nutritional needs of the population takes its place as one of the components of food and nutrition policy goals along with the priorities included in the FBDGs for improved health and nutrition for a given population.

The world nutrition and health situation demonstrates that the major causes of death and disability have been traditionally related to undernutrition in developing countries and to the imbalance between energy intake and expenditure (which lead to obesity and other chronic diseases - diabetes, cardiovascular disease, hypertension, and stroke) in industrialized countries. The tragedy is that many suffer from too little food while others have diseases resulting from too much food, but both would benefit from a more balanced distribution of food and other resources. Although the nature of the health and nutrition problems in these two contrasting groups is very different, the dietary guidelines required to improve both situations are not. Most countries presently have the combined burden of malnutrition from deficit and increasing prevalence of obesity and other chronic diseases from over consumption. The approaches to address the problems, nevertheless, should be country and population specific.

Although two-thirds of the world's population depends on cereal or tuber based diets, the other one-third consumes significant amounts of animal food products. The latter group places an undue demand on land, water, and other resources required for intensive food production, which makes the typical Western diet not only undesirable from the standpoint of health but also environmentally unsustainable. If we balance energy intake with the expenditure required for basal metabolism, physical activity, growth, and repair, we will find that the dietary quality required for health is essentially the same across population groups.

Efforts in nutrition education and health promotion should include a strong encouragement for active lifestyles. Improving energy balance for rural populations in developing countries may mean increasing energy intake to normalise low body mass index (BMI, weight/height^2, calculated as kg/m^2), ensuring adequate energy stores and energy for appropriate social interactions. In sedentary urban populations, improving energy balance will mean increasing physical activity to decrease energy stores (body fat mass) and thus normalise BMI. Thus, the apparent conflicting goals - eradicating undernutrition while preventing overnutrition - are resolved by promoting sufficient energy for a normal BMI. Moreover, if we accept that FBDGs should be ecologically sustainable, the types and amounts of foods included in a balanced diet are not very different for promoting adequate nutrition in the undernourished and preventing overnutrition in the affluent.

This is well exemplified by the similarities in the FBDGs across countries, whether represented by pyramids, rainbows, dishes, pots, etc. It is obvious that consumption of excess energy will induce an increase in energy stores, which may lead to obesity and related health complications. Populations should consume nutritionally adequate and varied diets, based primarily on foods of plant origin with small amounts of added flesh foods. Households should select predominantly plant-based diets rich in a variety of vegetables and fruits, pulses or legumes, and minimally processed starchy staple foods. The evidence that such diets will prevent or delay a significant proportion of non-communicable chronic diseases is consistent. A predominantly plant-based diet has a low energy density, which may protect against obesity. This should not exclude small amounts of animal foods, which may make an important nutritional contribution to plant-food-based diets, as illustrated in the examples presented earlier. Inadequate diets occur when food is scarce or when food traditions change rapidly, as is seen in societies undergoing demographic transitions or rapid urbanisation. Traditional diets, when adequate and varied, are likely to be generally healthful and more protective against chronic non-communicable diseases than the typical Western diet, consumed predominantly in industrialized societies.

Reorienting food production, agricultural research, and commercialisation policies needs to take into consideration FBDGs, which increase the demand for a variety of micronutrient-rich foods and thus stimulate production to meet the consumption needs. Prevailing agricultural policies encourage research on and production and importation of foods, which do not necessarily meet the requirements of FBDG implementation. For example, great emphasis is placed on cereals, horticultural crops for export, legumes for export, non-food cash crops, and large livestock. Necessary policy reorientation is required to ensure increased availability of micronutrient-rich foods within the local food system. Norway has successfully implemented agricultural and food production policies based on a National Nutrition Plan of Action, providing economic incentives for the producer and consumer in support of healthful diets. The results speak for themselves, as Norway has experienced a sustained improvement in life expectancy and a reduction in deaths from cardiovascular disease and other chronic non-communicable conditions.

Food Adulteration and Its Detection

Adulterated food is impure, unsafe, or unwholesome food. In the United States, the Food and Drug Administration (FDA), regulates and enforces laws on food safety and has technical definitions of adulterated food in various United States laws.

"Adulteration" is a legal term meaning that a food product fails to meet federal or state standards.

Federal Food, Drug, and Cosmetic Act, 1938

The Federal Food, Drug, and Cosmetic (FD&C) Act (1938) provides that food is "adulterated" if it meets any one of the following criteria: (1) it bears or contains any "poisonous or deleterioussubstance" which may render it injurious to health; (2) it bears or contains any added poisonous or added deleterious substance (other than a pesticide residue, food additive, color additive, or new animal drug, which are covered by separate provisions) that is unsafe; (3) its container is composed, in whole or in part, of any poisonous or deleterious substance which may render the contents injurious to health; or (4) it bears or contains a pesticide chemical residue that is unsafe. (Note: The United States Environmental Protection Agency (EPA) establishes tolerances forpesticide residues in foods, which are enforced by the FDA.)

Food also meets the definition of adulteration if: (5) it is, or it bears or contains, an unsafe food additive; (6) it is, or it bears or contains, an unsafe new animal drug;

(7) it is, or it bears or contains, an unsafe color additive; (8) it consists, in whole or in part, of "any filthy, putrid, or decomposed substance" or is otherwise unfit for food; or (9) it has been prepared, packed, or held underunsanitary conditions (insect, rodent, or bird infestation) whereby it may have become contaminated with filth or rendered injurious to health.

Further, food is considered adulterated if: (10) it has been irradiated and the irradiation processing was not done in conformity with a regulation permitting irradiation of the food in question (the FDA has approved irradiation of a number of foods, including refrigerated or frozen uncooked meat, fresh or frozen uncooked poultry, and seeds for sprouting [21 C.F.R. Part 179].); (11) it contains a dietary ingredient that presents a significant or unreasonable risk of illness or injury under the conditions of use recommended in labeling (for example, foods or dietary supplements containingaristolochic acids, which have been linked to kidney failure, have been banned.); (12) a valuable constituent has been omitted in whole or in part or replaced with another substance; damage or inferiority has been concealed in any manner; or a substance has been added to increase the product's bulk or weight, reduce its quality or strength, or make it appear of greater value than it is (this is "economic adulteration"); or (13) it is offered for import into the United States and is a food that has previously been refused admission, unless the person reoffering the food establishes that it is in compliance with U.S. law [21 U.S.C. § 342].

Poisonous or Deleterious Substances

Generally, if a food contains a poisonous or deleterious substance that may render it injurious to health. It can cause various harms. It is adulterated. For example, apple cider contaminated withE. coli O157:H7 and Brie cheese contaminated with Listeria monocytogenes are adulterated. There are two exceptions to this general rule. First, if the poisonous substance is inherent or naturally occurring and its quantity in the food does not ordinarily render it injurious to health, the food will not be considered adulterated. Thus, a food that contains a natural toxin at very low levels that would not ordinarily be harmful (for instance, small amounts of amygdalin in apricot kernels) is not adulterated.

Second, if the poisonous or deleterious substance is unavoidable and is within an established tolerance, regulatory limit, or action level, the food will not be deemed to be adulterated. Tolerances and regulatory limits are thresholds above which a food will be considered adulterated. They are binding on FDA, the food industry, and the courts. Action levels are limits at or above which FDA may regard food as adulterated. They are not binding on FDA. FDA has established numerous action levels (for example, one part per million methyl mercury in fish), which are set forth in its booklet Action Levels for Poisonous or Deleterious Substances in Human Food and Animal Feed.

If a food contains a poisonous substance in excess of a tolerance, regulatory limit, or action level, mixing it with "clean" food to reduce the level of contamination is not allowed. The deliberate mixing of adulterated food with good food renders the finished product adulterated (FDA, Compliance Policy Guide [CPG § 555.200]).

Filth and Foreign Matter

Filth and extraneous material include any objectionable substances in foods, such as foreign matter (for example, glass, metal, plastic, wood, stones, sand, cigarette butts), undesirable parts of the raw plant material (such as stems, pits in pitted olives, pieces of shell in canned oysters), and filth (namely, mold, rot, insect and rodent parts, excreta, decomposition). Under a strict reading of the FD&C Act, any amount of filth in a food would render it a, however, authorize the agency to issue Defect Action Levels (DALs) for natural, unavoidable defects that at low levels do not pose a human health hazard [21 C.F.R. § 110.110]. These DALs are advisory only; they do not have the force of law and do not bind FDA. DALs are set forth in FDA's Compliance Policy Guides and are compiled in the FDA and Center for Food Safety and Applied Nutrition (CFSAN) Defect Action Level Handbook.

In most cases, DALs are food-specific and defect-specific. For example, the DAL for insect fragments in peanut butter is an average of thirty or more insect fragments per 100 grams (g) [CPG § 570.300]. In the case of hard or sharp foreign objects, the DAL, which is based on the size of the object and the likelihood it will pose a risk of choking or injury, applies to all foods (see CPG § 555.425).

Economic Adulteration

A food is adulterated if it omits a valuable constituent or substitutes another substance, in whole or in part, for a valuable constituent (for instance, olive oil diluted with tea tree oil); conceals damage or inferiority in any manner (such as fresh fruit with food coloring on its surface to conceal defects); or any substance has been added to it or packed with it to increase its bulk or weight, reduce its quality or strength, or make it appear bigger or of greater value than it is (for example, scallops to which water has been added to make them heavier).

Microbiological Contamination and Adulteration

The fact that a food is contaminated with pathogens (harmful microorganisms such as bacteria, viruses, or protozoa) may, or may not, render it adulterated. Generally, for ready-to-eat foods, the presence of pathogens will render the food adulterated. For example, the presence of Salmonella on fresh fruits or vegetables or in ready-to-eat meat or poultry products (such as luncheon meats) will render those products adulterated.

For meat and poultry products, which are regulated by USDA, the rules are more complicated. Ready-to-eat meat and poultry products contaminated with pathogens, such as Salmonella or Listeria monocytogenes, are adulterated. (Note that hotdogs are considered ready-to-eat products.) For raw meat or poultry products, the presence of pathogens will not always render a product adulterated (because raw meat and poultry products are intended to be cooked, and proper cooking should kill pathogens). Raw poultry contaminated with Salmonella is not adulterated. However, USDA's Food Safety and Inspection Service (FSIS) has ruled that raw meat or poultry products contaminated with E. coli O157:H7 are adulterated. This is because normal cooking methods may not reduce E. coli

O157:H7 below infectious levels. E. coli O157:H7 is the only pathogen that is considered an adulterant when present in raw meat or poultry products.

Enforcement Actions against Adulterated Food

Ifafoodisadulterated,FDAandFSIShaveabroadarrayofenforcementtools.They are of various types. These include seizing and condemning the product, detaining imported product, enjoiningpersons from manufacturing or distributing the product, or requesting a recall of the product. Enforcement action is usually preceded by a Warning Letter from FDA to the manufacturer or distributor of the adulterated product. In the case of an adulterated meat or poultry product, FSIS has certain additional powers. FSIS may suspend or withdraw federal inspection of an official establishment. Without federal inspection, an establishment may not produce or process meat or poultry products, and therefore must cease operations. With the exception of infant formula, neither FDA nor FSIS has the authority to require a company to recall an adulterated food product. However, the ability to generate negative publicity gives them considerable powers of persuasion.

State regulators generally have similar enforcement tools at their disposal to prevent the manufacture and distribution of adulterated food. In addition, many states have the authority to immediatelyembargo adulterated food and to impose civil fines. Federal agencies often will coordinate with state or local authorities to remove unsafe food from the market as quickly as possible.

Food Adulteration Detection

Food Tea Leaves Adulterant Iron Flakes Detection Spread a small quantity (2 tea-spoon) of the sample on a piece of paper. Draw a magnet over it. Iron flakes, if present, cling to the magnet. The same test may be carried out to trace iron flakes from tea half-dust and iron filings from tea dust. Food Tea Leaves Adulterant Leather Flakes Detection Prepare a paper-ball. Fire the ball and drop a little amount of the sample on it. The presence of leather flakes emits an odour of burnt leather. Food Tea Leaves Adulterant Coal Tar Dye Detection Scatter a little amount (1 tea-spoon) of the sample on a moistened white blotting paper. After 5 minutes, remove the sample and examine the paper. A revelation of coloured spots indicates the use of the dye. Food Coriander powder and Cumin Powder Adulterant Saw Dust Detection Take a little amount (a half of tea-spoon) of the sample.

Sprinkle it on water in a bowl. Spice powder gets sedimented at the bottom and saw-dust floats on the surface. Food Green vegetables like Bitter Gourd, Green Chilli and others Adulterant Malachite Green Detection Take a small part of the sample and place it on a piece of moistened white blotting paper. The impression of colour on the paper indicates the use of malachite green, or any other low priced artificial colour. Food Arhar Pulse Adulterant Kesarri Pulse Detection Kesari Pulse has a characteristic wedge shape. Larger Kesari resembles Arhar (Tur). It can be separated by visual examination. Food Black Pepper Adulterant Papaya Seeds Detection Papaya seeds do not have any smell and are relatively smaller in size. Adulteration of papaya seed with Black Pepper may be detected by way of visual examination as also by way of smelling. Food Rice Adulterant Earth, sand, grit, unhusked paddy, rice bran, talc, etc. Detection These adulterants may be detected visually and removed by way of sorting, picking, and washing. Food Wheat Adulterant Earth, sand, grit, chopped straw, bran, unhusked grain, and seeds of weeds. Detection These adulterants may be detected visually and removed by way of sorting, picking, and washing.

Food Coffee powder Adulterant Cereal starch Detection Take a small quantity (one-fourth of a tea-spoon) of the sample in a test tube and add 3 ml of distilled water in it. Light a spirit lamp and heat the contents to colourize. Add 33 ml of a solution of potassium permanganate and muratic acid (1:1) to decolourize the mixture. The formation of blue colour in mixture by addition of a drop of 1% aqueous solution of iodine indicated adulteration with starch. Food Coffee powder Adulterant Powder of scorched persimmon stones Detection Take a small quantity (1 tea-spoon) of the sample and spread it on a moistened blotting paper. Pour on it, with much care, 3 ml of 2% aqueous solution of sodium carbonate. A red colouration indicates the presence of powder of scorched persimmon stones in coffee powder. Food Jaggery Adulterant Sodium bicarbonate Detection Take a little amount (one-fourth of a tea-spoon) of the sample in a test tube. Add 3 ml of muratic acid. The presence of sodium carbonate or sodium bicarbonate effects effervescence. Food Jaggery Adulterant Metanil yellow colour Detection Take a little amount (one-fourth of a tea-spoon) of the sample in a test tube. Add 3 ml of alcohol and shake the tube vigorously to mix up the contents. Pour 10 drops of hydrochloric acid in it. A pink colouration indicates the presence of metanil yellow colour in jaggery.

Food Asafoetida Adulterant Resin and colour Detection Take a little amount of small parts of the sample in test tube. Add 3 ml of distilled water and shake the tube gently. Pure asafoetida dissolves in water very quickly and produces a milky white colour, but in case of adulteration with a chemical colour the mixture turns to be coloured. The purity of asafoetida may also be examined by taking a little amount of it on the tip of a force and placing the same on the flame of a spirit lamp. Asafoetida burns quickly, producing bright flame and leaving the impurities behind.

Food Gram powder Adulterant Kesari powder Detection Take a little amount (a half of a tea-spoon) of the sample in a test tube with 3 ml of distilled water. Add 3 ml of muratic acid. Immerse the tube in warm water. Check the tube after 15 minutes. A violet colouration indicates the presence of Kesari powder in Gram powder.

Food Gram powder Adulterant Metanil yellow colour Detection Take a small quantity (a half of a tea-spoon) of the sample in a test tube. Add 3 ml of alcohol. Shake the tube to mix up the contents thoroughly. Add 10 drops of hydrochloric acid in it. A pink colouration indicates adulteration of gram powder with metanil yellow.

Food Processed food, sweetmeat or syrup Adulterant Rhodamine B colour Detection The presence of this chemical colour in food is very easy to detect as it shines very brightly under sun. A more precise methods of detection is also there. Take a little amount (a half of a tea-spoon) of the sample in a test tube. Add 3 ml of carbon tetrachloride and shake the tube to mix up the contents thoroughly. The mixture becomes colourless and an addition of a drop of hydrochloric acid brings the colour back when food contains Rhodamine B colour.

Food Processed food, sweetmeat or syrup Adulterant Metanil Yellow Detection Take little amount (a half of a tea-spoon) of the sample in a test tube. Add 10 drops of muratic acid or hydrochloric acid in it. The appearance of rosy colour indicates adulteration of food with metanil yellow.

Food Parched rice Adulterant Urea Detection Take 30 pieces of parched rice in a test tube. Add 5 ml of distilled water. Shake the tube to mix up the contents thoroughly. After 5 minutes, filter water contents and add to it a little amount (a half of a tea-spoon) of powder of arhar or soyabean. Wait for another 5 minutes and then dip a red litmus paper in the mixture. Lift the paper after 30 seconds and examine it. A blue colouration indicates the use of urea in parched rice. Food Turmeric powder Adulterant Metanil Yellow colour Detection Take a little amount (one-fourth of a tea-spoon) of the sample in a test tube. Add 3 ml of alcohol. Shake the tube to mix up the contents thoroughly. Add 10 drops of muratic acid or hydrochloric acid in it. A pink colouration indicates the use of metanil yellow colour in turmeric powder.

Food Green vegetable like green chilli etc. Adulterant Malachite Green Detection Rub the outer green surface of a small part of the sample with a liquid paraffin soaked cotton. The sample is adulterated when the white cotton turns green.adulterating causes severe diseases

Food Dry red chilli Adulterant Rhodamine B colour Detection Take a red chilli from the sample and rub the outer surface with a piece of cotton soaked in liquid paraffin. The sample is adulterated if the cotton becomes red.

Food Dry turmeric root Adulterant Metanil yellow colour Detection Take a piece of dry turmeric root and rub the outer surface with a piece of cotton soaked in liquid paraffin. A yellow colouration of cotton indicates adulteration of turmeric root with metanil yellow colour.

Food Sweet potato Adulterant Rhodamine B colour Detection Take a small part of the sample and rub the red outer surface with a piece of cotton soaked in liquid paraffin. The cotton adhering colour indicates the use of Rhodamine B colour on outer surface of the sweet potato.

Food is essential for nourishment and sustenance of life. Adulteration of food cheats the consumer and can pose serious risk to health in some cases. Food adulteration is thus a punishable offence under the provisions of Prevention of Food Adulteration Act, 1954 and the Rule made thereunder. Consumer awareness of quality of food and alertness in detecting common types of food adulteration can thus help in arresting this menace and enable Government to achieve the objective of Health for all by 2000 A. D.

Here we give the consumer an opportunity to detect a few common adulterants in food. These tests will simply help the consumers to screen their day-to-day food articles. But for detailed confirmatory test and quantification under P. F. A. Act the analysis in a laboratory is must.

Table 4.1 Common Adulterants and their Tests

Name of Food Article	Adulterant	Test for adulterants	Remarks
Milk	Water	The Lactometer reading shall not ordinarily be less than 26.	Lactometer is marked in degrees ranging from 0-40.
		The presence of water can be detected by putting a drop of milk on a polished slanting surface. The drop of pure milk either stops or flows slowly leaving a white trail behind it, whereas milk adulterated with water will flow immediately without leaving a mark.	The test is not valid if skimmed milk of other thickening material is added.
	Starch	Add a few drops of tincture of Iodine or Iodine solution. Formation of blue colour indicates the presence of starch.	
	Removal of Fat	The lactometer reading will go above 26 while the milk apparently remains thick.	
Khoa and it products	Starch	Boil a small quantity of sample with some water, cool and add a few drops of Iodine solution. Formation of blue colour indicates the presence of starch.	
Chhana or Paneer	Starch	Boil a small quantity of sample with some water, cool and add a few drops of Iodine solution. Formation of blue colour indicates the presence of starch.	

Name of Food Article	Adulterant	Test for adulterants	Remarks
Ghee and Butter	Vanaspati or Margarine	Take about one teaspoon full of melted sample of Ghee / Butter with equal quantity of concentrated Hydrochloric acid in a stoppered test tube and add to it a pinch of sugar. Shake well for one minute and let it stand for five minutes. Appearance of crimson colour in lower (acid) layer shows presence of Vanaspati or Margarine.	The test is specific for seasame oil which is compulsorily added to Vanaspati and Margarine. Some coal tar colours also gives a positive test. If the test is positive i.e red colour develops only by adding strong Hydrochloric acid (without adding crystals of sugar) then the sample is adulterated with coal tar dye.
	Mashed Potatoes, Sweet Potatoes and other starches	The presence of mashed potatoes and sweet potatoes in a sample of Ghee / Butter can easily be detected by adding a few drops of Iodine Solution. When Iodine, which is brownish in colour turns to blue then mashed potatoes / sweet potatoes or starches are present.	
Oils and Fats	Argemone oil	Take a small quantity of oil in a test tube. Add equal quantity of concentrated Nitric acid and shake carefully. Red to reddish brown colour in lower (acid) layer would indicate the presence of Argemone oil.	Colourless (not yellowish) Nitric acid may be used. Artificial colour if present will usually be a bright shade of colour, generally red or pink. The test may sometimes give misleading. The test may not respond if the Argemone oil is present in small quantity.

Name of Food Article	Adulterant	Test for adulterants	Remarks
	Mineral oil	Take 2 ml of the oil sample and add an equal quantity of N/2 Alcoholic potash. Heat in boiling water bath (dip in boiling water) for about 15 minutes and add 10 ml of water. Any turbidity shows presence of mineral oil.	If mineral oil is present in small quantity, this test may not be positive.
	Castor oil	Take about 1 ml of the oil, add 10 ml of acidified petroleum ether and mix well. Add a few drops of ammonium molybdate reagent. Immediate appearance of white turbidity indicates the presence of castor oil.	If castor oil is present in small quantity, this test may not be positive.
Sugar	Chalk powder	Dissolve 10 gm of sugar in a glass of water, allow it to settle. Chalk will settle down at the bottom.	
	Washing soda	Take small quantity of sugar, dissolve it in water and add few drops of Hydrochloric acid, effervescence (gives off bubbles) will indicate the presence of washing soda.	
Honey	Sugar solution	A cotton wick dipped in pure honey when lighted with a match stick burns and shows the purity of honey. If adulterated, the presence of water will not allow the honey to burn. If it does, it will produce a cracking sound.	This test is only for added water.

Name of Food Article	Adulterant	Test for adulterants	Remarks
Sweets, Ice-creams and beverages	Metanil yellow (a non permitted coal tar colour	Extract colour with luke-warm water from food articles. Add few drops of concentrated Hydrochloric acid. If magenta red colour develops the presence of metanil yellow is indicated.	
	Saccharin	Taste a small quantity. Saccharin leaves a lingering sweetness on tongue for a considerable time and leaves bitter taste at the end. Take two spoons of liquid sample or about 5 to 10 gms of solid sample with little quantity of water in a test tube and add few drops of Hydrochloric acid and 10 ml of solvent ether. Shake well. Decant the ether layer into a test tube or a beaker, evaporate the ether spontaneously. Add one drop of water (warm) to the residue and taste. Sweet taste will indicate the presence of saccharin.	See Appendix-II
	Aluminium foil	Aluminium foil is whitish grey in colour and is readily soluble in concentrated Hydrochloric acid while pure silver foil is not.	
Wheat, Rice, Maize, Jawar, Bajra, Chana, Barley, etc.	Dust, pebble, stone, straw, weed seeds, damaged grain, weevilled grain,	These may be examined visually to see foreign matter, damaged grains, discoloured grains, insect, rodent contamination etc.	Damaged / discoloured grains should be as low as possible since they may be affected by fungal toxins, Argemone seeds, Dhatura seeds etc.

Name of Food Article	Adulterant	Test for adulterants	Remarks
	insects, rodent hair and excreta.		In moderately excessive amount can result in risk to health. Discard the damaged / undesirable grains before use.
	Ergot (a fungus containing poisonous substance)	Purple black longer sized grains in Bajra shows the presence of Ergots. Put some grains in a glass tumbler containing 20 percent salt solution. Ergot floats over the surface while sound grains settle down.	
	Dhatura	Dhatura seeds are flat with edges with blackish brown colour which can be separated out by close examination.	
	Karnel Bunt	The affected wheat karnel have a dull appearance, blackish in colour and rotten fish smell.	
	Argemone seed	Assemble mustard seed but shows a protrusion on close examination. The surface of argemone seed is grainy and rough while that of mustard seed is smooth. When mustard seed is pressed, inside it is yellow whereas argemone seed is white.	
Parboiled Rice	Metanil yellow	Rub a few grains in the palms, yellow colour would get reduced or disappear. Add a few drops of dilute Hydrochloric acid to a few rice grains mixed with little water, presence of pink colour indicates, presence of Metanil yellow.	

Name of Food Article	Adulterant	Test for adulterants	Remarks
	Turmeric	Take a small amount of sample in a test tube, add some water and shake. Dip Boric acid paper (filter paper dipped in Boric acid solution), if it turns pink, turmeric is present.	See Appendix-I
Dal whole and spilt	Khesari Dal	Add 50 ml of dilute Hydrochloric acid to the sample and keep on simmering water for about 15 minutes. The pink colour developed indicates the presence of khesari dal.	The test is only for khesari dal. (Metanil yellow if presents will give a similar colour immediately even without simmering)
	Clay, stone, gravels, webs, insects, rodent hair and excreta	Visual examination will detect these adulterants.	Reject if the number of insects is large or if the odour is unpleasant and taste bitter or gritty.
	Metanil yellow	Take 5 gms of the sample with 5 ml. of water in a test tube and add a few drops of concentrated Hydrochloric acid. A pink colour shows the presence of Metanil yellow.	
Atta, Maida, Suji (Rawa)	Sand, soil, insects, webs, lumps, rodent hair and excreta	These can be identified by visual examination.	
	Iron fillings	By moving a magnet through the sample, iron fillings can be separated.	

Name of Food Article	Adulterant	Test for adulterants	Remarks
Besan	Khesari flour	Add 50 ml of dilute Hydrochloric acid to 10 gms of the sample and keep on simmering water for about 15 minutes. The pink colour developed indicates the presence of khesari flour.	
Whole spices	Dirt, dust, straw, insect, damaged seeds, other seeds, rodent hair and excreta	These can be examined visually.	
Black pepper	Papaya seeds	Papaya seeds can be separated out from pepper as they are shrunken, oval in shape and greenish brown or brownish black in colour.	
	Light black pepper	Float the sample of black pepper in alcohol. The mature black pepper berries sink while the papaya seeds and light black pepper float.	
	Coated with mineral oil	Black pepper coated with mineral oil gives Kerosene like smell.	
Cloves	Volatile oil extrated (exhausted cloves)	Exhausted cloves can be identified by its small size and shrunken appearance. The characteristic pungent taste of genuine cloves is less pronounced in exhausted cloves.	

Name of Food Article	Adulterant	Test for adulterants	Remarks
Mustard seed	Argemone seed	Mustard seeds have a smooth surface. The argemone seed have grainy and rough surface and is black and hence can be separated out by close examination. When mustard seed is pressed, inside it is yellow while for argemone seed it is white.	Use magnifying glass for identification.
Powdered spices	Added starch	Add a few drops of tincture of Iodine or Iodine solution. Formation of blue colour indicates the presence of starch.	Iodine test for added starch is not applicable for turmeric powder.
	Common salt	Taste for addition of common salt.	
Turmeric powder	Coloured saw dust	Take a teaspoon full of turmeric powder in a test tube. Add a few drops of concentrated Hydrochloric acid. Instant appearance of pink colour which disappear on dilution with water shows the presence of pure turmeric. If the colour persists, metanil yellow is present.	This test is only for Metanil yellow.
	Chalk powder or yellow soap stone powder	Take a small quantity of turmeric powder in a test tube containing small quantity of water. Add a few drops of concentrated Hydrochloric acid, effervescence (give off bubbles) will indicate the presence of chalk or yellow soap stone powder.	

Name of Food Article	Adulterant	Test for adulterants	Remarks
Chillies powder	Bricks powder, salt powder or talc powder	Take a teaspoon full of chillies powder in a glass of water. Coloured water extract will show the presence of artificial colour. Any grittiness that may be felt on rubbing the sediment at the bottom of glass confirms the presence of brick powder / sand, soapy and smooth touch of the white residue at the bottom indicates the presence of soap stone.	This test is only for earthy materials.
	Water soluble coal tar colour	Water soluble artificial colour can be detected by sprinkling a small quantity of chillies or turmeric powder on the surface of water contained in a glass tumbler. The water soluble colour will immediately start descending in colour streaks.	
	Oil soluble coal tar colour	Take 2 gms of the sample in a test tube, add few ml of solvent ether and shake. Decant ether layer into a test tube containing 2 ml of dilute Hydrochloric acid (1 ml HCL plus 1 ml of water). Shake it, the lower acid layer will be coloured distinct pink to red indicating presence of oil soluble colour.	See also Appendix-I
Hing	Soap stone or other earthy matter	Shake little portion of the sample with water and allow it to settle. Soap stone or other earthy matter will settle down at the bottom.	In compounded Hing due to presence of starch a slight turbid solution may be produced. However, this will settle down after keeping.

Name of Food Article	Adulterant	Test for adulterants	Remarks
Saffron	Dried tendrils of maize cob	Genuine saffron will not break easily like artificial. Artificial saffron is prepared by soaking maize cob in sugar and colouring it with coal tar colour. The colour dissolves in water if artificially coloured. A bit of pure saffron when allowed to dissolved in water will continue to give its saffron colour so long as it lasts.	
Common salt	White powdered stone	Stir a spoonful of sample of salt in a glass of water. The presence of chalk will make solution white and other insoluble impurities will settle down.	
Tea leaves	Exhausted tea or tur or gram dal husk with colour	Take a filter paper and spread a few tea leaves. Sprinkle with water to wet the filter paper. If coal tar colour is present it would immediately stain the filter paper. Wash the filter paper under tap water and observe the stains against light. Spread a little shaked lime on procelain tile or glass plate. Sprinkle a little tea dust on the lime. Red, orange or other shades of colour spreading on the lime will show the presence of coal tar colour. In case of genuine tea, there will be only a slight greenish yellow colour due to chlorophyl, which appear after sometime.	

Name of Food Article	Adulterant	Test for adulterants	Remarks
	Iron fillings	By moving a magnet through the sample, iron fillings can be separated.	
Coffee	Chicory	Gently sprinkle the coffee powder sample on the surface of water in a glass. The coffee floats over the water but chicory begins to sink down within a few seconds. The falling chicory powder particles leave behind them a trail of colour, due to large amount of caramel.	
	Tamarind seeds powder and Date seeds powder	Sprinkle the suspected coffee powder on white filter / blotting paper and spray 1 percent sodium carbonate solution on it. Tamarind and date seed powder will stain filter / blotting paper red.	
Pan Masala	Colour and Saccharin	Colour dissolves in water. Saccharin gives excessive and lingering sweet taste and leaves bitter taste at the end.	
Catachu powder	Chalk	Chalk gives effervescence (gives off bubbles) with concentrated Hydrochloric acid.	This test is only for chalk.

Name of Food Article	Adulterant	Test for adulterants	Remarks
Silver leaves	Aluminium leaves	On ignition, genuine silver leaves burn away completely, leaving glistering white spherical ball of the same mass whereas aluminium leaves are reduced to ashes of dark grey blackish colour. Take silver leaves in test tube, add diluted Hydrochloric acid. Appearance of turbidity to white precipitate indicates the presence of silver leaves. Aluminium leaves do not give any turbidity or precipitate. Take a small portion of metal leaves and add a few drops of concentrated Nitric acid. Silver leaves will completely dissolve whereas aluminium leaves will remain undissolved.	
Vinegar	Mineral acid	Test with the Metanil yellow indicator paper. In case, the colour changes from yellow to pink, mineral acid is present.	See Appendix-I

Method for Test

1. *Test for Metanil Yellow:* Take some sample in a test tube and add some amount of water, shake well. Add few drops of diluted Hydrochloric acid. Violet colour in the water portion indicates the presence of Metanil Yellow.
2. *Test for Starch:* Boil the sample with some water in a test tube, cool and add a few drops of Iodine solution. Appearance of blue colour indicates the presence of Starch.
3. *Baudouin test:* Take about one teaspoon full of melted Ghee / Butter with equal quantity of concentrated Hydrochloric acid in a test tube and add a pinch of sugar to it. Shake well and allow to stand. Appearance of crimson red colour shows the presence of Vanaspati or Margarine.
4. *Boric acid test for Turmeric:* Take a small amount of sample in a test tube, add some water and shake. Dip Boric acid paper. If it turns pink, turmeric is present. (Boric acid paper can be prepared by dipping a strip of filter paper in the Boric acid solution. Boric acid solution can be prepared by dissolving 5 gms of Boric acid in 100 ml concentrated Hydrochloric acid.)
5. *Metanil yellow indicator paper:* Metanil yellow indicator paper can be prepared by dipping a strip of filter paper in Metanil yellow solution (1 gm Metanil yellow coal tar colour dissolved in 100 ml of water).
6. *Oil soluble coal tar colour:* Take a small quantity of chillies powder in a beaker and add 5 ml of rectified sprit (alcohol). Dip a small piece of white silk cloth for two minutes. Remove the silk cloth piece and wash with water. If the silk cloth is permanently dyed, it indicates the presence of oil soluble coal tar colour.

List of Apparatus and Reagents for Developing a Simple Kit

Apparatus

1. Magnifying Glass
2. Spatula
3. Magnet
4. Forcep
5. Lactometer
6. Beaker
7. Dishes
8. Dropper
9. Reagent Bottles
10. Spirit lamp
11. Test tube ordinary
12. Test tube stoppered
13. Glass rod
14. Test tube stand
15. Small plastic tray (White)
16. Porcelain tile (White)
17. Glass Cylinder
18. Glass Marking pen
19. Filter paper
20. White silk cloth

21. Cotton

Reagents

1. Hydrochloric acid
2. Nitric acid
3. Petroleum ether
4. Solvent ether
5. Rectified spirit
6. Iodine / Tincture of Iodine
7. Potassium Hydroxide
8. Ammonium Molybdate
9. Boric Acid
10. Sodium Carbonate
11. Metanil yellow powder

Caution

1. The testing kit should be kept beyond the reach of children as it contains harmful chemicals.
2. Solvent ether is highly inflamable. Keep it away from fire. Don't keep the bottle's cap open.
3. Acids are highly corrosive. In case of acid burn wash immediately with cold water containing sodium bicarbonate.
4. Use gloves while performing the tests.

Dietary Supplement

A dietary supplement is intended to provide nutrients that may otherwise not be consumed in sufficient quantities.

Supplements as generally understood include vitamins, minerals, fiber, fatty acids, or amino acids, among other substances. U.S. authorities define dietary supplements as foods, while elsewhere they may be classified as drugs or other products.

There are more than 50,000 dietary supplements available. More than half of the U.S. adult population (53% - 55%) consume dietary supplements with most common ones being multivitamins These products are not intended to prevent or treat any disease and in some circumstances are dangerous, according to the U.S. National Institutes of Health. For those who fail to consume a balanced diet, the agency says that certain supplements "may have value."[3] Effects of most of these products have not been determined in randomized clinical trials and manufacturing is lightly regulated; randomized clinical trials of certain vitamins and antioxidants have found increased mortality rates.

Need for Dietary Supplements

Dietary supplements are unnecessary if one eats a balanced diet. Moreover, many supplements have no practical effect. For example, glucosamine and chondroitin, often thought to relieve joint pain, have been shown to be without benefit. Relief reported for these supplements may be explained by the fact that many people taking them on a doctor's recommendation are also taking other pain relievers. Another common supplement, omega-3 fatty acids, has been similarly shown to be without benefit for healthy individuals.

Some supplements are harmful. For example, one study funded by the National Institute of Health found that men who consumed vitamin E supplements had a higher rate of prostate cancer than men who did not. Also, it has been found that calcium supplements puts people at a greater risk for heart attacks.

Supplements may create harm in several ways, including over-consumption, particularly of minerals and fat-soluble vitamins which can build up in the body. The products may also cause harm related to their rapid absorption in a short period of time, quality issues such as contamination, or by adverse interactions with other foods and medications.

Bodybuilding Supplement

Bodybuilding supplements are dietary supplements commonly used by those involved in bodybuilding and athletics. Bodybuilding supplements may be used to replace meals, enhance weight gain, promote weight loss or improve athletic performance. Among the most widely used are vitamin supplements, protein, branched-chain amino acids (BCAA), glutamine, essential fatty acids, meal replacement products, creatine, weight loss products and testosterone boosters. Supplements are sold either as single ingredient preparations or in the form of "stacks" - proprietary blends of various supplements marketed as offering synergistic advantages. While many bodybuilding supplements are also consumed by the general public their salience and frequency of use may differ when used specifically by bodybuilders.

Annual sales of sport nutrition products in the US is over $2.7 billion (US) according to *Consumer Reports*.

Athletes in ancient Greece were advised to consume large quantities of meat and wine. A number of herbal concoctions and tonics have been used by strongmen and athletes since ancient times across cultures for the enhancement of strength, vigour, prowess and stamina. In the 1910s, Eugen Sandow, widely considered to be the first modern bodybuilder in the West, advocated the use of dietary control to enhance muscle growth. Later, bodybuilder Earle Liederman advocated the use of "beef juice" or "beef extract" (basically, consomme) as a way to enhance muscle recovery. In 1950s with recreational and competitive bodybuilding becoming increasingly popular Irvin P. Johnson began to popularize and market egg-based protein powders marketed specifically at bodybuilders and physical athletes. The 1970s and 1980s marked an explosion in the growth of the bodybuilding supplement industry fueled by an unprecedented increase in mainstream recreational bodybuilding and the widespread use of modern marketing techniques.

In the USA, in October 1994, the Dietary Supplement Health and Education Act (DSHEA) was signed into law. Under DSHEA, a supplement manufacturer alone is responsible for determining that the dietary supplements it manufactures or distributes are safe. Dietary supplements did not henceforth need approval from FDA before they were marketed. Except in the case of a new dietary ingredient, a firm did not have to provide FDA with the evidence to substantiate safety or effectiveness. It is widely believed that the 1994 DSHEA further consolidated the position of the supplement industry and lead to unprecedented growth and sales figures.

Supplement Categories

Modern bodybuilding supplements are often marketed as promoting various desirable processes related to improving nutrition, enhancing body composition or improving lifting performance. Supplements are often categorized accordingly. While many of these categories are based on scientifically based physiological or biochemical processes, their use in bodybuilding parlance is often heavily colored by bodybuilding lore and industry marketing and as such may be deviate considerably from traditional scientific usages of these terms.

Protein shake

Bodybuilders often supplement their diets with a powdered form of protein. The powder is mixed with water, milk or juice. Protein powder is generally consumed immediately before and after exercising, or in place of a meal. Some types of protein are to be taken directly before and after a workout (whey protein), while others are to be taken before going to bed (casein protein). The theory behind this supplementation is that bodybuilders, by virtue of their unique training methods and end-goals, require higher-than-average quantities of protein to support maximal muscle growth.

Currently, no consensus has been reached in determining whether or not an individual in exercise training can benefit from protein and amino acid supplements. Protein supplements come in various forms: ready to drink shakes, bars, bites, oats, gels and powders. Protein powders are available in a variety of flavors.

Whey protein contains high levels of all the essential amino acids and branched-chain amino acids. It also has the highest content of the amino acidcysteine, which aids in the biosynthesis of glutathione. For bodybuilders whey protein provides amino acids used to aid in muscle recovery.[4] Whey protein is derived from the process of making cheese from milk. There are three types of whey protein: whey concentrate, whey isolate, and whey hydrolysate. Whey concentrate is 29–89% protein by weight whereas whey isolate is 90%+ protein by weight. Whey hydrolysate is enzymatically predigested and therefore has the highest rate of digestion of all protein types.

- Casein protein (or milk protein) has glutamine, and casomorphin.
- Soy protein from soybeans contain isoflavones, a type of phytoestrogen.
- Egg-white protein is a lactose- and dairy-free protein.
- Hemp seed contains complete and highly-digestible protein and hemp oil is high in essential fatty acids.
- Rice protein, when made from the whole grain, is a complete protein source that is highly digestible and allergen free. Since rice protein is low in the amino acid lysine, it is often combined with pea protein powder to achieve a superior amino acid profile.

- Pea protein is a hypoallergenic protein with a lighter texture than most other protein powders. Pea protein has an amino acid profile similar to that of soy, but pea protein does not elicit concerns about unknown effects of phytoestrogens. Pea protein is also less allergenic than soy.

Although it is generally undisputed that athletes and bodybuilders need an increased intake of protein, the exact amount is highly individualized and dependent on the type and duration of the exercise as well as the physiological make up of the individual. Age, gender, and body size may vary this protein intake. Some health experts have criticized protein shakes as being unnecessary for most people that consume them, since most users already get enough protein in the normal varied diet with enough calories.[6] However, there is some evidence to support the idea that protein shakes are superior to whole foods with regards to enhancing muscle hypertrophy in the one hour window following intensive exercise. Moreover, for athletes who do not have the time to prepare whole food meals on the run or immediately after exercise, a protein shake may be preferred for practical as well as performance reasons.

Additionally, some studies suggest low-calorie dieters, vegetarians, haphazard eaters and those who train very heavily may benefit significantly from protein supplements. Traditional nutrition theory states that the body can only metabolize 5–9 grams of protein per hour, and that excessive daily intake can cause weight gain, kidney problems, osteoporosis, or diarrhea. However, many bodybuilders report consuming hundreds of grams of protein per day to achieve maximal strength gains, so this nutrition requirement may only apply to the general public. Taking an overdose of protein can lead to a loss of appetite, which may be useful for some dieters. Nutritionists claim that osteoporosis occurs from excessive protein intake because protein can put pressure on the kidneys and lead to bone loss due to calcium leaching. However, recent research has cast doubts on these claims, and suggests that higher calcium excretion may be due to increased calcium absorption in the intestines due to protein intake. Indeed, it is well known that dietary protein is itself important for bone growth, and some studies have found increased bone formation in response to exchanging dietary carbohydrates for protein. Nutritionists also argue against increased protein consumption because weight gain may occur because, as the body cannot store protein, excess protein will either be burned as energy or stored as fat (if you are already getting the calories you need). However, dietary protein is converted to fat far less efficiently than either carbohydrates or lipids, so consuming a calorie excess in protein will result in far less fat gain that would a calorie excess of other macronutrients.[7] Research by Tarnopolsky*et al.* (1988) showed that for bodybuilding individuals, 1.90 g of protein per kg of body weight per day is recommended, whereas endurance athletes require 1.34g/kg/d of protein. Their findings indicated that protein requirements are actually much lower than might be expected and that protein supplements therefore may not be as effective as is popularly believed.

However, it should be noted that both of these levels are significantly higher than the levels claimed to be necessary for the general population (0.8 g protein

/ kg body weight). Studies suggest that there are different protein requirements for anaerobic and aerobic exercise. Endurance athletes in aerobic activity may have increased daily protein intake at 1.2–1.4 g per kg body weight per day where strength training athletes performing anaerobic activity may have increased daily protein intake needs at 1.4–1.8 g per kg body weight so as to enhance muscle protein synthesis or to make up for the loss of amino acid oxidation during exercise.

Branched-chain amino acids

Amino acids are the building blocks of protein; the body breaks consumed protein into amino acids in the stomach and intestines. There are three branched-chain amino acids (BCAAs): leucine,isoleucine, and valine. Each has numerous benefits on various biological processes in the body. Unlike other amino acids, BCAAs are metabolised in the muscle and have an anabolic/anti-catabolic effect on it. BCAAs account for 33% of muscle protein.

Glutamine

Glutamine is the most abundant amino acid found in human muscle and is commonly found in supplements or as a micronized, instantly soluble powder because supplement manufacturers claim the body's natural glutamine stores are depleted during anaerobic exercise. Serum glutamine is used by the body to counteract the acidosis that results from exercise; in order to replenish the loss of glutamine from the bloodstream, the body catabolizes glutamine from the muscle. Ingestion of supplemental glutamine may therefore help ensure a ready supply for the muscles. It is also argued that a deficiency may lead to a weakened immune system and wasting of muscle tissue.

Some studies have shown there to be no significant effect of glutamine on bench press strength, knee-extension torque or lean muscle mass when compared to controls taking a placebo, though another study found that glutamine is beneficial in raising T-helper/suppressor cell ratio in long-distance runners.

Essential fatty acids

The essential fatty acids (alpha-linolenic acid and linoleic acid) may be especially important to supplement while bodybuilding; these cannot readily be made in the body, but are required for various functions within the body to take place.

Fatty fish, such as fresh salmon and trout are rich in essential fatty acids and fish oils can also be taken in supplement form.

Flaxseed oil, often sold as a supplement on its own, is an ideal source of alpha-Linolenic acid, which can also be found in walnuts and pumpkin seeds.

Meal Replacement Products

Meal replacement products (MRPs) are either pre-packaged powdered drink mixes or edible bars designed to replace prepared meals. MRPs are generally high in protein, low in fat, have a low to moderate amount of carbohydrates, and contain a wide array of vitamins and minerals.

The majority of MRPs use whey protein, casein (often listed as calcium caseinate or micellar casein), soy protein, and/or egg albumin as protein sources. Carbohydrates are typically derived frommaltodextrin, oat fiber, brown rice, and/or wheat flour. Some MRPs also contain flax oil powder as a source of essential fatty acids.

MRPs can also contain other ingredients, such as creatine monohydrate, glutamine peptides, L-glutamine, calcium alpha-ketoglutarate, additional amino acids, lactoferrin, conjugated linoleic acid, and medium-chain triglycerides.

A sub-class of MRPs are called 'weight gainers' and have a high ratio of carbohydrates:protein. Where a MRP would have a 0.25-2:1 ratio of Carbohydrates:Protein a weight gainer would have in the order of between 3-5:1 ratios.

Prohormones

Prohormones are precursors to hormones and were most typically sold to bodybuilders as a precursor to the natural hormone testosterone. This conversion requires naturally occurring enzymes in the body. Side effects are not uncommon, as prohormones can also convert further into DHT and estrogen. To deal with this, many supplements also have aromatase inhibitors and DHT blockers such as chrysin and 4-androstene-3,6,17-trione. To date most prohormone products have not been thoroughly studied, and the health effects of prolonged use are unknown. Although initially available over the counter, their purchase was made illegal without a prescription in the US in 2004, and they hold similar status in many other countries. Additionally, their use is proscribed by most sporting bodies.

Creatine

Creatine is an organic acid naturally occurring in the body that supplies energy to muscle cells for short bursts of energy (as required in lifting weights) via creatine phosphate replenishment ofATP. A number of scientific studies have shown that creatine can improve strength, energy, muscle mass, and recovery times. In addition, recent studies have also shown that creatine improves brain function.[22] and reduces mental fatigue. Unlike steroids or other performance-enhancing drugs, creatine can be found naturally in many common foods such as herring, tuna, salmon, and beef.

Creatine increases what is known as *cell volumization* by drawing water into muscle cells, making them larger. This *intracellular* retention should not be confused with the common myth that creatine causes bloating (or *intercellular* water retention).

Creatine is sold in a variety of forms, including creatine monohydrate and creatine ethyl ester, amongst others. Though all types of creatine are sold for the same purposes, there are subtle differences between them, such as price and necessary dosage.

In *The New Encyclopedia of Modern Bodybuilding, 2nd ed.,* author Arnold Schwarzenegger states:

Creatine monohydrate is regarded as a necessity by most bodybuilders. Creatine monohydrate is the most cost-effective dietary supplement in terms of muscle size and strength gains. ... There is no preferred creatine supplement, but it is believed that creatine works best when it is consumed with simple carbohydrates. This can be accomplished by mixing powdered creatine with grape juice, lemonade, or many high glycemic index drinks.

Some studies have suggested that consumption of creatine with protein and carbohydrates can have a greater effect than creatine combined with either protein or carbohydrates alone.

Thermogenic Products

A thermogenic is a broad term for any supplement that the manufacturer claims will cause thermogenesis, resulting in increased body temperature, increased metabolic rate, and consequently an increased rate in the burning of body fat and weight loss. Until 2004 almost every product found in this supplement category comprised the "ECA stack": ephedrine, caffeine and aspirin. However, on February 6, 2004 the Food and Drug Administration (FDA) banned the sale of ephedra and its alkaloid, ephedrine, for use in weight loss formulas. Several manufacturers replaced the ephedra component of the "ECA" stack with bitter orange or citrus aurantium (containing synephrine) instead of the ephedrine.

Testosterone Boosters

There are several naturally occurring plants, amino acids, and vitamins, as well as synthetic chemicals that are sold as supplements to increase testosterone levels. Some commonly taken supplements of this type are Fenugreek, Eurycoma longifolia, D-Aspartic acid, Boron, L-Carnitine and Tribulus terrestris.

Food Fortification

Foodfortificationorenrichmentistheprocessofaddingmicronutrients(essential trace elements and vitamins) to food. It can be purely a commercial choice to provide extra nutrients in a food, or sometimes it is a public health policy which aims to reduce numbers of people with dietary deficiencies in a population.

Diets that lack variety can be deficient in certain nutrients. Sometimes the staple foods of a region can lack particular nutrients, due to the soil of a region, or because of the inherent inadequacy of the normal diet. Addition of micronutrients to staples and condiments can prevent large-scale deficiency diseases in these cases.

While it is true that both fortification and enrichment refer to the addition of nutrients to food, the true definitions do slightly vary. As defined by the World Health Organization (WHO) and theFood and Agricultural Organization of the United Nations (FAO), fortification refers to "the practice of deliberately increasing the content of an essential micronutrient, ie. vitamins and minerals (including trace elements) in a food irrespective of whether the nutrients were originally in the food before processing or not, so as to improve the nutritional quality of the food supply and to provide a public health benefit with minimal risk to health," whereas enrichment is defined as "synonymous with fortification and refers to the addition of micronutrients to a food which are lost during processing."

Food fortification was identified as the second strategy of four by the WHO and FAO to begin decreasing the incidence of nutrient deficiencies at the global level.

As outlined by the FAO, the most common fortified foods are:

- Cereals and cereal based products
- Milk and Milk products
- Fats and oils
- Accessory food items
- Tea and other beverages
- Infant formulas

Types of Food Fortification

The four main methods of food fortification (named as to indicate the procedure that is used in order to fortify the food):

1. Biofortification (i.e. breeding crops to increase their nutritional value, which can include both conventional selective breeding, and modern genetic modification)thetic biology (i.e. addition of probiotic bacteria to foods)
2. Commercial and industrial fortification (i.e. flour, rice, oils (common cooking foods))
3. Home fortification (e.g. vitamin D drops)[4]

Rationale

The WHO and FAO, among many other nationally recognized organizations, have recognized that there are over 2 billion people worldwide who suffer from a variety of micronutrient deficiencies. In 1992, 159 countries pledged at the FAO/WHO International Conference on Nutrition to make efforts to help combat these issues of micronutrient deficiencies, highlighting the importance of decreasing the number of those with iodine, vitamin A, and iron deficiencies.[2] A significant statistic that led to these efforts was the discovery that approximately 1 in 3 people worldwide were at risk for either an iodine, vitamin A, or iron deficiency. Although it is recognized that food fortification alone will not combat this deficiency, it is a step towards reducing the prevalence of these deficiencies and their associated health conditions.

In Canada, The Food and Drug Regulations have outlined specific criterion which justifies food fortification:

1. To replace nutrients which were lost during manufacturing of the product (i.e. the manufacturing of flour
2. To act as a public health intervention
3. To ensure the nutritional equivalence of substitute foods (i.e. to make butter and margarine similar in content, soy milk and cow's milk, etc.)
4. To ensure the appropriate vitamin and mineral nutrient composition of foods for special dietary purposes (i.e. Boost, gluten-free products, low sodium, or any other products specifically designed for special dietary requirements from an individual).[7]

There are also several advantages to approaching nutrient deficiencies among populations via food fortification as opposed to other methods. These may include, but are not limited to: treating a population without specific dietary interventions therefore not requiring a change in dietary patterns, continuous delivery of the nutrient, does not require individual compliance, and potential to maintain nutrient stores more efficiently if consumed on a regular basis.

Chapter 5

Determination of Fat Content

The methods used in determination of fat content in foods are as varied as the sample matrices. This is an overview of a few methods applicable or adaptable to a wide variety of foods. The various fat determination methods lead to different definitions of total fat content. For the purpose of labeling, the FDA has defined total fat as the fatty acids aobtained from a total lipid extract of the food matrix expressed as triglycerides. These methods of extraction and/or detection may or may not adhere strictly to this definition, and this issue may need to be investigated further depending on the requirements of the determination.

Infrared Spectrometry Method

In this method the fat is extracted from a homogeneous sample by a chloroform-methanol solvent containing an internal standard. The internal standard found to be most suitable for this procedure was a low viscosity apolar methylsilicone oil (MS 550). The extraction is then analyzed by infrared spectrometry, where the carbonyl ester band of the fat is at 5.72 μm and the internal standard of choice is at 7.88 μm. The areas under the two bands are determined by the triangulation method, or by integration with the IR spectrometer if available. The band area ratio (esterified fatty acids/internal standard) and a calibration curve of weight of fat versus the band area ratio are used to determine the fat content. The calibration curves are prepared by using this same procedure to analyze samples of pure reference fat in the range including the expected fat content of the analyte.

The accuracy and precision of this method are comparable to reference methods which included Rose-Gottlieb for ice cream and liquid egg, Gerber for milk, and Werner-Schmid for other samples. The accuracy in 15 of 20 sample products was not significantly different at the 5% level. For the 5 samples, the fat content was as much as 17% different. Precision varied depending on the food analyzed. The average absolute standard deviation for all products was 0.22 for the reference method and 0.28 for the IR method.

Although the limit of detection was not specifically addressed, this study was conducted on samples ranging from 3.3% to 32.54% fat content. IR analysis showed good comparison with standard reference methods in most samples. Observed variations related to food matrix rather than percent fat content.

The sensitivity of this method is quite good as the slope of the calibration curves range from 60 to 80 mg fat. The calibration curve is the graph of milligrams of fat versus the band area ratio. The selectivity of this method is better than the reference methods. While the reference methods typically estimate all lipid material in a product, this method determines only the total esterified fatty acids resulting in a lower fat content determination in many products. This agrees with the FDA lableing requirement.

For linear dynamic range considerations, the calibration curves demonstrate very good linearity in the band area ratio range of 0.9-1.4, and deviate only at band area ratios exceeding 3.5.

Overall this method provides a rapid method of fat determination which shows good accuracy and precision depending on the analyte. All the figures of merit are acceptable. If the IR instrument does not have integration capabilities, the tedious triangulation method could be prohibitively slow. Although the calibration curve in some cases depends on the type of food analyzed, most foods can be analyzed with a composite calibration curve.

Caviezel Method

The homogenized sample and the internal standard (IS, tridecanoic acid) are added to the n-butyl alcohol solvent. Potassium hydroxide is used to saponify and extract the fats simultaneously. An acidic aqueous solution is added to convert the fatty acids salts to fatty acids, producing a two phase system where the fats and internal standard are contained in the top layer. This is injected into the fat determination system where it is separated by gas chromatography. The peak areas of the IS and fatty acids are used to determine the fat content which is then converted to triglyceride content with a predetermined factor. The fat determination unit is a commercial instrument which is based on gas chromatography and is geared specifically to this analysis method. This company also produces mixers and extraction units which are designed for this analysis.

The accuracy of this method is very good. A regression of the measured fat content versus the certified fat content of several standard reference materials (SRMs, pork meat, baby food powder, rye flour, and chocolate) resulted in a slope of 0.990 and R squared of 99.89 demonstrating a strong correlation.

Reproducibility and repeatability were tested with nine types of samples with fat contents ranging from 1 to 79%, taking ten replicates on ten different days. The repeatability relative standard deviations values, RSDr were 0.47 to 3.59% while the reproducibility relative standard deviation values, RSDR, ranged from 0.85 to 9.52%. The recovery from spiked samples averaged 97.2% and yielded a linear regression with R squared 99.96. The concentration range studied was 0.3 to 100% and the regression of measured versus added fat was linear with R squared of 99.99.

Limits of detection showed 4.3 mg absolute fat in 5 g corresponding to 0.1% fat by the AOAC and ACS definition of limit of detections as blank + 3 standard deviations. According the ACS definition of limit of quantitation as blank signal + 10 standard deviations, this method gives 8.2 mg absolute fat or 0.2% fat content.

The laboratories in this study described this method as well documented, easier to follow and shorter than comparable methods. Other comments included that this method involved easier sample handling, and that the instrument was easy to operate. The cost of the gas chromatography based detection unit was not quoted.

Simplified Gravimetric after Chloroform-Methanol Extraction Method

Samples of 1-5 g are massed into a polypropylene centrifuge bottle. Sodium acetate is added so that the total volume of solution is 32 ml. Aliquots of methanol and chloroform are added, the bottles capped and shaken for 2 hours. Another aliquot of chloroform is added, and the bottles shaken again for 30 minutes. Finally, an aliquot of water is added, and the bottles are shaken a final time for 30 min. Centrifuge tubes which had been dried and massed are used to hold 20 ml aliquots of the chloroform layer. These are then centrifuged for 10 min, and allowed to set in 25DC water bath for 15 minutes.

The samples are then evaporated to dryness under a nitrogen blanket, heated in a drying oven for 30 minutes, and cooled in a dessicooler for at least 30 minutes. Finally, they are massed and the total lipid content determined by: total lipid (g/100 g wet weight) = (W2-W1)*Vc*100/(VA*SW) where W2 is the weight of glass tube and dried extract (g), W1 is the weight of empty dried glass tube (g), Vc is the total volume of chloroform (ml), VA is the volume of extract dried (ml), and SW is the weight of food sample assayed (g).

The accuracy of the method was determined by results for NIST Standard reference material (SRM) 1548. The mean determined was 20.9 ± 0.52 g/100g dry wt., well within the certified range of 20.6 ± 2 g/100g dry weight.

Recovery was essentially quantitative in spiked samples of typical foods (3.2% by weight total lipid) to which ranges of 0.1 to 1.0 g fat was added to 5 g aliquots. The concentration range is limited to 3 to 30% total weight of fat. The precision was determined with 608 samples with average total lipid content from 2.4 to 8.7 g /100 g wet weight. The within sample covariation was 0 to 9%.

The selectivity can be a concern as the chloroform-methanol extraction may overestimate fat content because it extracts lipids other than fatty acids. On the other hand, in may underestimate fat content because low molecular fatty acids might not be extracted.

This is a modification of AOAC 983.23 and involves less analyst time, less solvent loss and the Clarase reagent is not needed. It is more applicable to batch analysis and simpler than the AOAC method, yet the unfavorable halogenated hydrocarbon solvent is still required. Gravimetric analyses are severely dependant on mass and volume accuracy. This factor leaves this method generally unfavorable.

Supercritical CO2 Extraction Method

The finely ground sample is massed and placed in an extraction cartridge and mixed with an equal amount of ISCO (ISCO Inc, Lincoln, NE) wet support. SFC/SFE grade CO2 is used with a commercial instrument at pressures up to 10,000 psi, temperatures of about 80DC and approximately a 4 mL/min flow of liquid CO2 for about 30 minutes to extract fat content. The extraction parameters vary depending on the sample matrix. After extraction the collection vial is heated for 20 minutes at 120DC. The collection vial is massed before and after extraction. The variable conditions include extractor temperature, pressure, modifiers of ethanol or ethanol and water mixture, restrictor temperature, flow rate, and extraction time.

Standard figures of merit were not reported in this article. SFE was compared to standard reference methods for six different categories of foods, including dairy, chocolate, and gravies. The reference methods were International Dairy Federation, Weibull, Twisselmann, and several IUPAC methods. In all cases SFE showed bias in reporting a lower fat content than the reference. The repeatability and reproducibility in most cases were good, however not quite as good as the reference methods. The concentration range was from less than 1% to 35% fat content.

Overall this method provides a rapid, fully automated determination. There is a significant reduction in the used of solvents and avoids the use of the halogenated hydrocarbons, there is low thermal damage to the sample and there exists the possibility of selectively isolating components. The high cost and complex equipment are the main disadvantages. There is also incomplete recovery for some products, as well as the fact that the method and conditions must be developed for different sample matrices.

The methods reviewed all had acceptable figures of merit. Supercritical CO2 extraction is the simplest and quickest method, however, the capital equipment cost may be prohibitive. The Caviezel method is also rapid and largely automated, but again the cost of special equipment may be undesirable. If a large capital expenditure and method development expense is acceptable, the supercritical CO2 method is advantageous in the elimination of the use of halogenated hydrocarbons. The gravimetric method is unfavorable due to the errors introduceable in the mass and volume measurements. Although the IR method uses the unfavorable halogenated hydrocarbon, it has the advantage of some automation with the use of nonspecialized equipment. Assuming an IR spectrometer with integration capabilities is available, this method would be preferred.

Chapter 6

Determination of Protein Nitrogen

Cereals and seeds are significant components of the human diet and the principal part of feeding stock for domestic animals. One of the most important nutrients of cereals and seeds is protein. The monitoring of protein amount, through the estimation of nitrogen, must be accurate in order to determine the nutritional quality of cereals and seeds. If the amount of nitrogen is multiplied by a factor depending on the kinds of protein expected to be present in the food, then the total protein content can be determined. In addition to its dietary importance, the protein content has also become a guideline for some cereal trade transactions.

This article discusses current legislation governing the analysis of nitrogen content for protein determination in cereals and seeds and examines the use of an advanced combustion technique against the traditional Kjeldahl method. The advantages of the combustion method are further highlighted with an application example.

The Kjeldahl Method

This is a method for the quantitative determination of nitrogen in chemical substances. For many years, the Kjeldahl method has been the internationally-recognised method for estimating protein concentration in food, due to its universality, precision and reproducibility.

Currently, it is the standard method against which all other methods are judged. Nevertheless, the Kjeldahl method is associated with many major downfalls. It is not capable of providing measurements of true protein content and different correction factors are required for different proteins to account for different amino acid sequences. The Kjeldahl method is a multi-stepped process, including sample digestion in boiling sulphuric acid, neutralisation with sodium hydroxide solution, distillation of the resulting ammonia gas into a trapping solution, titration with an acid solution and determination of the amount of nitrogen and protein by calculation. The entire process may require over four hours to be completed while all steps, excluding digestion, require continuous technician contribution.

In addition, handling boiling sulphuric acid, especially with the addition of concentrated caustic solution, is a particularly hazardous task. Kjeldahl analyses also generate toxic waste since they involve the use of mercury or selenium catalysts during the digestion step.

All these disadvantages have ended the dominance of the Kjedahl method as the preferred technique for nitrogen/protein determination in food. On the contrary, the capabilities of the combustion method have been greatly improved to make it faster, safer and more reliable than the traditional Kjeldahl method.

Designed to offer an alternative to the Kjeldahl methodology for dedicated nitrogen and nitrogen/protein analysis in food, the upgraded combustion method features a large sample load of 2g, ensuring maximum accuracy and reliability of results as well as improved productivity capabilities, reducing sample handling time and minimising matrix effects. The technique reduces the need for sample preparation and is capable of processing more samples than the Kjeldahl method in only a few minutes. In addition, consumable costs associated with sample preparation are minimised and safety in the laboratory is increased.

The advanced combustion method features a unique automatic carbon dioxide (CO2) adsorber regenerating technology, which is automatically activated to adsorb the CO2 generated during the combustion.

Unlike traditional technologies, this self-cleaning filter never needs to be changed, saving time and money. This feature improves the autonomy of the system and reduces instrument downtime and maintenance. The benefits of the combustion method are further strengthened by an inherent water condensation drainage device (Peliter) and the combination of a safe, sensitive and reliable thermal conductivity detector (TCD), featuring an extensive working range to cover nitrogen/protein applications from low ppm to high percentage concentrations.

Applications

Since 1883, the Kjeldahl method has gained wide acceptance and is now used for a variety of applications. Kjeldahl nitrogen determinations are performed on food and beverages, meat, feed, grain, waste water, soil and many other samples. The method has been refined and tested for a wide variety of substances and approved by various scientific associations including:

- AOAC International (formerly the Association of Official Analytical Chemists)
- Association of American Cereal Chemists
- American Oil Chemists Society
- Environmental Protection Agency

- International Standards Organization
- United States Department of Agriculture

What is the Kjeldahl method?

The Kjeldahl method is a means of determining the nitrogen content of organic and inorganic substances. Although the technique and apparatus have been altered considerably over the past 100 years, the basic principles introduced by Johan Kjeldahl endure today.

The Kjeldahl method may be broken down into three main steps:

- Digestion: the decomposition of nitrogen in organic samples utilizing a concentrated acid solution. This is accomplished by boiling a homogeneous sample in concentrated sulfuric acid. The end result is an ammonium sulfate solution.
- Distillation: adding excess base to the acid digestion mixture to convert NH4+ to NH3, followed by boiling and condensation of the NH3 gas in a receiving solution.
- Titration: to quantify the amount of ammonia in the receiving solution.

The amount of nitrogen in a sample can be calculated from the quantified amount of ammonia ions in the receiving solution.

The Digestion Process

A general equation for the digestion of an organic sample is shown below as one basic example:

Organic N + H2SO4 . (NH4)2SO4 + H2O + CO2 + other sample matrix by-products A number of interrelated digestion conditions determine the rate of reaction and the completeness of the breakdown of nitrogen to ammonium sulfate. Among these are heat input to the acid digestion mixture, amount of inorganic salt added to elevate the acid boiling temperature, reflux rate of H2SO4 in the neck of the digestion flask, length of digestion, and catalyst addition. Adjusting any one of these factors has an influence on the others. Proper digestion conditions for a given sample matrix are achieved through establishing a balance of these factors in a controlled and repeatable fashion. In addition, if the sample contains nitrate or nitrite nitrogen, it is possible to chemically pretreat the digest to include or exclude this nitrogen source from the analysis as desired in a particular situation.

Acid Considerations

Sulfuric acid has been used alone for the digestion of organic samples. (As a convenience to reduce bumping of the digestion mixture, Alundum boiling chips or pumice are often added.) The amount of acid required is influenced by sample size and relative amount of carbon and hydrogen in the sample, as well as amount of nitrogen. A very fatty sample consumes more acid. Also, heat input and digestion length influences the amount of acid loss due to vaporization during the digestion process. Remember that a Kjeldahl flask is essentially a bulb with a condenser neck off to the side for refluxing of the acid.

Heat Input and Digestion Length

Typically the heating elements used for Kjeldahl digestions have variable settings. Heat input is frequently specified as "that setting which brings 250 ml of water at 25° C to a rolling boil in 5 minutes."

Initially an organic sample usually chars and blackens. The reaction may at first be very vigorous depending on the matrix and the heat input. With organic decomposition the digestion mixture gradually clears as CO_2 evolves. Metallic ions might tint the clear digestion mixture. Note that solution clearing in itself is not an indication that all organic nitrogen has been broken down. Digestion length must be determined by recovery studies on known materials of similar matrix if a new method is being developed.

Salt Additions

The problem with using sulfuric acid alone for digestion is very long digestion times result with many samples due to the slow rate of organic decomposition. The addition of an inorganic salt to the digest elevates the boiling point of the H_2SO_4. The solution temperature of concentrated sulfuric acid alone is about 330° C. Addition of a salt such as K_2SO_4 can elevate the solution temperature of the digestion mixture to 390° C or more, depending on the ratio of salt to acid. This significantly increases the rate of organic decomposition in the digestion mixture, shortening the length of time required for digestion.

There are several precautions to keep in mind concerning salt addition. First, it is possible to raise the solution temperature of the digestion mixture too much. If the temperature goes much above 400° C during any phase of the digestion, volatile nitrogen compounds may be lost to the atmosphere.

Remember that as acid is gradually consumed during the digestion process, for the various reasons mentioned above, the salt acid ratio of the digest gradually rises. This means that the hottest solution temperatures are attained at the end of the digestion. Heat input, consumption of acid by organic material and vaporization, salt/acid ratio, digestion length, and physical design of the Kjeldahl flask, are all interrelated. Each has an effect on the final solution temperature.

A second precaution is that if the salt/acid ratio is too high, a considerable amount of material will "salt out" upon cooling of the digest. Concentrated acid pockets can be contained within the cake. These can react violently when concentrated base is added in the distillation process. A certain amount of salting out can be managed by diluting the digest with water while it is still somewhat warm, but not too hot.

Catalyst Additions

Several catalysts have been employed by Kjeldahl chemists over the years to increase the rate of organic breakdown during the acid digestion. Mercuric oxide has been the most effective and widely used. However, mercury forms a complex with ammonium ions during digestion. The addition of sodium thiosulfate or sodium sulfide after digestion and before distillation will break the complex and precipitate mercuric sulfide. This is also important from a safety point of view, as mercury vapor might escape to the atmosphere during the distillation process.

Because of environmental concerns over the handling and disposal of mercury, other catalysts are coming more into favor. Many methods employ copper sulfate. Titanium oxide and copper sulfate in combination have been found to be more effective than copper sulfate alone. Selenium is frequently used. Commercially prepared mixtures of potassium sulfate and a catalyst are available from laboratory chemical suppliers. Bulk custom mixtures are also available.

Nitrate and Nitrite Reduction

Kjeldahl digestions do not always recover all forms of nitrogen in a sample. Nitrate and nitrite ions in a sample must first be reduced prior to acid digestion for quantitative recovery. Salicylic acid followed by sodium thiosulfate has been used to pretreat the mixture to ensure complete reduction. Other reduction schemes have been devised. Or other pretreatments have been used to prevent nitrates from being reduced at all during the charring process, leaving a clear digest with no contribution from nitrate ions.

The Distillation Process

The acid digestion mixture is diluted and made strongly alkaline with NaOH, liberating NH3 as follows:

Ammonium ammonia sulfate heat gas

(NH4)2SO4 + 2NaOH . 2NH3. + Na2SO4 + 2H2O

The Kjeldahl flask is attached to a water condenser and is heated to boil off the NH3 gas from the digest. The tip of the condenser is submerged in a flask of acidic receiving solution, either standard acid or boric acid solution, to again trap the distilled NH3 in receiving solution.

Digestion Mixture Dilution

The acid digestion mixture is usually cooled and diluted with ammonia-free water. As mentioned above, with digestion mixtures containing high salt/acid ratios, dilution prevents or minimizes caking. Sometimes this is done while the digestion mixture is still quite warm (with caution!), and the K2SO4 has not yet salted out. With some digestion mixtures, a cake forms, but dissolves or breaks up if time is allowed following dilution. Alternatively, warming or sonicating after dilution may break up the cake. Cake material may cause low nitrogen recoveries in the distillation step. Also, entrapped acid in a cake may react violently with base during the distillation process.

Dilution of the digestion mixture before making it alkaline and distilling also reduces the likelihood of bumping. Additional boiling chips added just before distillation also reduces bumping, especially towards the end of the distillation as the solution becomes more concentrated. Two or three drops of tributyl citrate may be added as an aid to reduce foaming.

NaOH Addition

Concentrated NaOH (usually 50% solution) is added slowly down the neck of the flask. Being heavier, it forms a layer underneath the diluted acid digestion mixture. Generally, for each 5 ml of concentrated sulfuric acid used in the digestion, 20 ml of 50% sodium hydroxide is required to make the digest strongly alkaline (pH of >11). The flask is connected to the condenser and mixed before heating and distillation begins.

For samples not requiring a digestion step, such as direct ammonia determinations in water, the sample is buffered to a pH of 9.5 with a solution of sodium tetraborate and sodium hydroxide, to decrease hydrolysis of any complex organic nitrogen compounds present.

Distillation

The majority of the NH3 is distilled and trapped in the receiving acid solution within the first 5 or 10 minutes of boiling. But depending on the volume of the digestion mixture and the method being followed, 15 to 150 ml of condensate should be collected in the receiving flask to ensure complete recovery of nitrogen. Further extension of the distillation times and volumes collected simply results in more water being carried over to the receiving solution. Excess water does not change the titration results.

Distillation times and distillate volumes collected should be standardized for all samples of a given methodology. The rate of distillation is affected by condenser cooling capacity and cooling water temperature, but primarily by heat input. Typically the heating elements used for distillation have variable temperature controllers. A distillation rate of about 7.5 ml/minute is most commonly cited in accepted methods.

Connecting bulbs or expansion chambers between the digestion flask and the condenser is an important consideration to prevent carryover of the alkaline digestion mixture into the receiving flask. The slightest bit of contamination of the receiving solution can cause significant error in the titration step.

When very low levels of nitrogen are being determined, it is advisable to "precondition" the distillation apparatus prior to distillation. This can be done by distilling a 1:1 mixture of ammonia-free water and 50% NaOH for 5 minutes just before sample distillation to reduce contamination from atmospheric ammonia.

Receiving Solutions

If the receiving solution is standardized HCl or H2SO4, it is desirable to have only a slight excess left after the NH3 is distilled and trapped in the receiving solution to minimize the back titration. Based on the anticipated amount of nitrogen in the sample, a target amount of standard acid can be calculated from the following formula:

If boric acid is used, the exact concentration is not needed because the titration directly measures the amount of ammonia in the distillate by neutralizing the 1:1 complex formed by ammonia and boric acid. Large quantities of boric acid may be added to the receiving solution so complete absorption of the ammonia is assured.

Receiving solution volumes may be increased by the addition of ammonia-free water so that the tip of the delivery tube is immersed. Delivery tubes should always be allowed to drain momentarily into the receiving flask before removal from the distillation apparatus. The receiving solution should remain below 45° C during distillation to prevent loss of ammonia.

The Titration Process

There are two types of titration: back titration, commonly used in Macro Kjeldahl; and direct titration. Both methods indicate the ammonia present in the distillate with a color change and allow for calculation of unknown concentrations.

Nitrogen Determination by Back Titration

The ammonia is captured by a carefully measured excess of a standardized acid solution in the receiving flask. The excess of acid in the receiving solution keeps the pH low, and the indicator does not change. The excess acid solution is exactly neutralized by a carefully measured standardized alkaline base solution such as sodium hydroxide. A color change is produced at the end point of the titration.

Nitrogen Determination by Direct Titration

If boric acid is used as the receiving solution instead of a standardized mineral acid, the chemical reaction is:

excess ammonia boric ammonium-borate boric gas acid complex acid

$NH_3 + H_3BO_3 . NH_4^+ :H_2BO_3^- + H_3BO_3$

The boric acid captures the ammonia gas, forming an ammonium-borate complex. As the ammonia collects, the color of the receiving solution changes.

ammonium sulfuric ammonium boric borate complex acid sulfate acid -$2NH_4H_2BO_3 + H_2SO_4 . (NH_4)_2SO_4 + 2H_3BO_3$

The addition of sulfuric acid exactly neutralizes the ammonium borate complex, and a reverse color change is produced. The boric acid method has two advantages: only one standard solution is necessary for the determination and the solution has a long shelf life.

Indicator Solutions

Many different indicators have been used to provide a "sharp end point" color change. The analyst's use of specific types of indicators can be a personal choice.

The combination of methyl red and methylene blue indicators is frequently used in many methods. The color change of the indicator must fall within the equivalent point of the reaction. For standard acid/base titrations, methyl orange is usually the preferred indicator. If color change end points are difficult to detect, reference solutions made from a blank with an indicator can be very helpful.

Calculations

The calculations for % nitrogen or % protein must take into account which type of receiving solution was used and any dilution factors used during the distillation process. The equations given here are in long form. They are often simplified in the published standard methods. In the equations below, "N" represents normality. "ml blank" refers to the milliliters of base needed to back titrate a reagent blank if standard acid is the receiving solution, or refers to milliliters of standard acid needed to titrate a reagent blank if boric acid is the receiving solution. When standard acid is used as the receiving solution, the equation is:

[(ml standard acid x N of acid) - (ml blank x N of base)] -(ml std base x N of base) x 1.4007

weight of sample in grams

If the sample weight is in milligrams, the molecular weight of nitrogen should be changed to 1400.67.

When boric acid is used as the receiving solution the equation is:

(ml standard acid - ml blank) x N of acid x 1.4007 weight of sample in grams

If it is desired to determine % protein instead of % nitrogen, the calculated % N is multiplied by a factor, the magnitude of the factor depending on the sample matrix. Many protein factors have been developed for use with various types of samples. The list below represents just a few of the factors described in the standard methods of analysis published by the American Association of Cereal Chemists (AACC) and AOAC International.

6.38 milk and dairy

6.25 other grains

5.95 rice

5.70 wheat flour

Equipment and Apparatus

Very early Kjeldahl digestions and distillations were performed using stone fume hoods and gas mantles as a heat source. In the 1920's these were replaced by what is now known as classical macro-Kjeldahl digestion and distillation apparatus. Macro setups use Kjeldahl flasks from 500 to 800 ml volume and handle sample sizes from 0.5 to about 5.0 g.

A smaller version of this apparatus is referred to as micro-Kjeldahl equipment. The equipment consists of smaller benchtop heater units, and Kjeldahl flasks of 30 to 100 ml volume.

A third more recent variation in equipment makes use of ceramic or aluminum heating blocks designed to accept a number of straight digestion tubes at once. "Block digestors" are often used in conjunction with benchtop distillation units with steam generators to shorten the distillation time.

Each of these three types of apparatus are described in more detail below. In all cases, since the Kjeldahl process involves significant corrosive fumes, appropriate attention must be given to fume removal. Equipment must be suitably constructed of corrosion-resistant materials.

Classical Macro-Kjeldahl Apparatus

Many governmental and regulating agencies have developed methodologies that specify the classical macro-Kjeldahl apparatus. For example, standard methods for low level nitrogen determinations in water (0-10 mg/1) require a sample size of 250 to 500 ml, and therefore large Kjeldahl flasks. Methodologies in the agricultural industry often require larger sample sizes and involve matrices which can significantly foam.

Equipment engineered to accommodate macro methods can be substantial. Apparatus designed to process 2 to 12 samples simultaneously is typical, and involves a bank of heaters and a fume manifold for digestion, and a second bank of heaters with condensers for distillation. The apparatus is available freestanding or contained within an integral hood. Freestanding installations require additional laboratory room modifications to handle fumes and heat. In either case the equipment is large, expensive, involves significant installation considerations, and ongoing utility and reagent costs.

The digestion apparatus has separate heating mantles to retain each flask. Typically each mantle has an individual temperature controller. The long necks of the Kjeldahl flasks are positioned to release digestion fumes into a common manifold. Manifold aspiration can be provided either by a mechanical blower with exhaust through ductwork to the outside, or by a large water aspirator that provides a water spray to dilute the fumes for disposal down a drain.

Like the digestion apparatus, the distillation apparatus has separate heating mantles to retain each flask. With the flask in place on the mantle, the neck of each flask is attached to a "connecting bulb" or an expansion chamber that acts as a trap to keep any of the concentrated liquid digest from mechanically carrying over through the condensers and into the receiving solution.

Downstream from the connecting bulb is the condenser, which is a tube of stainless steel surrounded by a second water jacket tube. Typically multiple condenser/jacket units are ganged together in a single assembly. A glass delivery tube is attached to the end of the condenser. The delivery tube has a ball-shaped tip with small holes to help disperse large bubbles and eliminate pressure fluctuations during distillation. The tip is submerged in the flask of receiving solution to ensure that the distilling ammonia is completely captured.

Micro-Kjeldahl Apparatus

These are miniature versions of the macro-digestion apparatus which are movable and intended for use in a laboratory hood. They are designed to digest small samples in 30 or 100 ml digestion flasks. Individually controlled heaters allow multiple flasks to be handled simultaneously. Some micro-digestors include a one-piece glass manifold for fume removal. The glass manifold is connected to a standard water aspirator to dilute and remove fumes that do not reflux in the flask necks. Whether or not a glass manifold is used, the digestion apparatus should be operated inside a laboratory hood.

Digested samples can be diluted and an aliquot taken for distillation on a micro steam distillation unit. Micro steam distillation units are available to accept digested samples of up to 4 ml concentrated acid and volumes of about 55 ml. An electric immersion heater produces steam heat that causes the sample to boil and release ammonia gas. The vapors pass into a condenser where contact with water-cooled glass condenses the vapors, which drip into the receiving solution via a delivery tube. Distillation time is approximately 5 minutes.

Methodologies have been specifically developed for micro-Kjeldahl apparatus. These most often are methods that involve homogeneous samples, relatively high nitrogen levels in the sample, and small sample sizes normally less than 0.25 g.

The installation considerations, initial expense, space requirements, and ongoing utility and reagent cost are much less with micro equipment.

Block Digestors

Block digestors are high temperature ceramic or aluminum blocks with wells to accept straight walled digestion tubes. They can accommodate from 2 to 25 tubes at once. Block digestors are heavy but movable self-contained units designed to be used in a laboratory hood. Digestion tubes range up to 300 ml capacity, suitable for sample sizes up to about 2 g or 75 ml.

Since narrow, straight-walled digestion tubes are used instead of traditional Kjeldahl flasks with a bulb and neck, and since a number of tubes are grouped closely together in a vertical configuration in the block, acid reflux parameters are different. This means salt/acid ratios and heat inputs appropriate for macro-Kjeldahl methods do not exactly transfer to digestion in a block. Methods have been developed specifically for block digestors similar, but not exactly identical, to macro-Kjeldahl methods.

Typically block digestors have one controller that adjusts the temperature of the entire block. The controller can be automated, allowing for timed ramping and multiple temperature settings during the course of a digestion.

Most often block digestors are used together with fume removal devices, even though they are operated in a hood. One such system consists of a glass manifold with individual bulb eductors in a rack designed to rest on top of a set of digestion tubes in the block. Another system uses individual exhaust caps with flexible tubing. In either case the manifold is connected to a water aspirator. Or, instead of a water aspirator, there are alternative units that employ mechanical aspiration and include bubbling the acid fumes through a neutralizing base solution. Block digestion followed by steam distillation is sometimes referred to as "rapid Kjeldahl" partly because a number of block digestion methodologies are shorter than classical macro-Kjeldahl digestions and benchtop steam distillation units in use are faster than classical distillations.

Rapid Steam Distillation Apparatus

The development of block digestors as an alternative to macro-Kjeldahl digestion has been paralleled by the development of benchtop steam distillation units. Basically, this equipment consists of a steam generator to inject steam into the alkaline digestion mixture and an evaporator-type condenser, where contact with water-cooled glass coils causes the vapors to condense and drip into the receiving solution via a delivery tube. Usually the unit is designed to accept straight digestion tubes from block digestors, though there is no reason in theory why any digestion techniques could not be combined with any distillation technique. Steam distillation is much more rapid than classical macro-distillation, typically taking from 3 to 8 minutes. On the other hand, only one distillation at a time can be done.

Rapid steam distillation units are available with different features, ranging from relatively manual models to highly automated models. Basic models dispense dilution water and base under control of a push button. Some models include a timer to control the length of distillation, shutting down automatically. Some units automate the entire distillation process once the digestion flask is in place, and others automatically titrate to end point after distillation, and calculate and display the results in a printed report.

Titrators

Of course, receiving solutions may be individually hand titrated using an indicator solution and buret, but several models of benchtop automatic titrator instruments are also available. Some units will titrate one receiving solution at a time to a set end point. Others will automatically titrate a number of receiving flasks sequentially. Still other models provide a printout of results along with digital readout.

Other Equipment and Procedures

Following Kjeldahl digestion, there are other procedures available to determine nitrogen (or protein) in the digestion mixture. The most popular involves the spectrophotometric method for ammonium determinations of digestion mixtures. The Nessler and Phenate color development methods are the most widely used. The Biuret method, a dye binding procedure, is another type. Automated analyzers which use color development chemistry are available and can perform large numbers of assays. However, the apparatus and maintenance costs are high.

Manual Nessler colorimetric methods are often used in ammonia determinations. Digested samples are prepared for a color development and their absorbency is read with a spectrophotometer at a range of 400 to 500 nm. Standards are also run and a standard curve is prepared which plots absorbency versus milligrams of ammonia. The standard curve is then compared to the sample results.

The conditions required for optimal color development and stability can be rigorous for all of these procedures. However, spectrophotometric analysis has the ability to determine very small quantities of ammonia — as low as 20 micro-grams per liter under optimal conditions.

Ammonia specific ion electrodes give the analyst another means for the determination of ammonia. The electrode uses a hydrophobic gas permeable membrane to separate the sample solution from an internal ammonium chloride solution. Dissolved ammonia diffuses through the membrane and changes the internal pH, which is sensed by a pH meter that has an expanded millivolt scale. Ammonia concentrations of standards versus potential in millivolts are plotted on a semi-logarithmic graph, and sample concentrations are read from the standard curve. The ammonia specific ion electrode can be precise and has few interferences.

In addition, there is alternative nitrogen determination equipment which involves techniques other than Kjeldahl digestion. There are instruments which incorporate techniques of Dumas combustion, neutron activation, near infrared reflectance spectrometry or direct distillation. The first two instrument/techniques determine total nitrogen regardless of its form within the sample. Their acceptance in most laboratories has been limited due to the necessary changes involved in reporting these types of nitrogen values. These changes would involve broad multi-industrial re-evaluations of comparative nitrogen-protein analysis.

Near infrared reflectance spectrometry and direct distillation use methods based on approximate relationships between nitrogen content and protein. Both methods depend on calibrations with standards which have been determined by traditional Kjeldahl analysis.

Most Common Problems

Like any other method, the Kjeldahl procedure is subject to problems with methodology, technique and equipment. Understanding the principles behind the Kjeldahl process and using appropriate quality control procedures should prevent most problems from occurring. When developing a new or modifying an existing method, one should proceed with caution and experimental verification.

Should nitrogen results differ from those expected, knowledge of common problems makes finding and correcting the error easier. Common problems include the following: sample size and type incompatibility with amounts and types of acid, salt or catalyst used; contaminated samples, standards, reagents or equipment; inadequate or excessive digestion time; inadequate or excessive distillation dilution volume or aliquot volume; foaming or bumping during digestion; uneven digestion; inadequate or excessive sample temperature;

precipitation of salt or caking or digested sample; leaks in apparatus connections; inadequate sodium hydroxide addition or sodium hydroxide carryover; and calculation errors.

Legislation

As a direct consequence of these key advantages, the combustion method has been approved and adopted by the Association of Official Analytical Chemists (AOAC) and the American Association of Cereal Chemists (AACC).

The official methodology requirements for protein determination by combustion include AOAC Method 992.23 and AACC Method 46-30, 1999. These methods indicate that the suitable fineness of grind must be determined for each different material analysed to achieve precision that gives RSD (equal or less) of two per cent for ten successive determinations of nitrogen. The AACC 46-30 crude protein combustion method specifies that nitrogen, freed by pyrolysis and subsequent combustion at high temperature in pure oxygen, is quantified by thermal conductivity detection and converted to equivalent protein by appropriate numerical factor. This generic combustion method is applicable to all flours, cereal grains, oilseeds and animal feeds. Different conversion factors are used for various cereal grains and oilseeds. The AOAC 990.03 protein in animal feed combustion method is applicable to the determination of nitrogen in all types of forages. Providing nutrition information on food labels is important to public health as it helps consumers to adopt a balanced diet. Current European legislation mandates that only foods claiming a specific nutrition value, for example 'low fat' or 'high in fibre', such as cereals, should carry nutritional information. These products must label energy content, seven core nutrients including protein, carbohydrates, total fat, saturated fat, trans fat, sodium and sugars and any nutrient for which a claim is made. This regulatory requirement further stresses the need for a powerful protein monitoring method.

Experimental

A Thermo Scientific FLASH 4000 nitrogen/protein analyser was used to validate the combustion method in terms of accuracy and reproducibility according with the pre-treatment of the sample. Different cereals and seeds were chosen for this analysis. Protein content was calculated automatically using the Thermo Scientific Eager Xperience software with default protein factor of 6.25 (5.70 for rice). The protein factor can be changed in accordance with the food type.

Temperature of the left reactor was set at 950°C and 840°C for the right reactor. Temperature of the oven was 50°C. A 500 mg EDTA (9.59 per cent nitrogen) standard was used and a sample weight of 600mg-1.6g. The oxygen amount necessary for the complete combustion of samples was calculated automatically by the OxyTune function present in the Eager Xperience software. Food safety is a growing and important market that requires superior analytical accuracy in order to ensure optimum productivity. The conventional Kjeldahl method for nitrogen determination is extremely time consuming, requiring extensive sample preparation and handling and using various unsafe chemicals and solutions. The Kjeldahl method is also associated with high waste costs and is incapable of operating continuously while being strongly dependent on the user's experience. As a consequence, productivity is considerably jeopardised.

However, recent advances of the combustion method have made this technique perfectly suitable to cope effortlessly with the wide array of food safety laboratory requirements such as accuracy, day-to-day reproducibility and high sample throughput.

Chapter 7

Determination of Moisture

Moisture content is one of the most commonly measured properties of food materials. It is important to food scientists for a number of different reasons:

- *Legal and Labeling Requirements.* There are legal limits to the maximum or minimum amount of water that must be present in certain types of food.
- *Economic.* The cost of many foods depends on the amount of water they contain - water is an inexpensive ingredient, and manufacturers often try to incorporate as much as possible in a food, without exceeding some maximum legal requirement.
- *Microbial Stability.* The propensity of microorganisms to grow in foods depends on their water content. For this reason many foods are dried below some critical moisture content.
- *Food Quality.* The texture, taste, appearance and stability of foods depends on the amount of water they contain.
- *Food Processing Operations.* A knowledge of the moisture content is often necessary to predict the behavior of foods during processing, *e.g.* mixing, drying, flow through a pipe or packaging.

It is therefore important for food scientists to be able to reliably measure moisture contents. A number of analytical techniques have been developed for this purpose, which vary in their accuracy, cost, speed, sensitivity, specificity, ease of operation, *etc*. The choice of an analytical procedure for a particular application depends on the nature of the food being analyzed and the reason the information is needed.

Properties of Water in Foods

The moisture content of a food material is defined through the following equation:

$$\%\text{Moisture} = (m_w/m_{sample})\square 100$$

Where m_w is the mass of the water and m_{sample} is the mass of the sample. The mass of water is related to the number of water molecules (n_W) by the following expression: $m_w = n_w M_w/N_A$, where M_w is the molecular weight of water (18.0 g per mole) and N_A is Avadagro's number (6.02□10^{23} molecules per mole). In principle, the moisture content of a food can therefore be determined accurately by measuring the number or mass of water molecules present in a known mass of sample. It is not possible to directly measure the number of water molecules present in a sample because of the huge number of molecules involved. A number of analytical techniques commonly used to determine the moisture content of foods are based on determinations of the mass of water present in a known mass of sample. Nevertheless, as we will see later, there are a number of practical problems associated with these techniques that make highly accurate determinations of moisture content difficult or that limit their use for certain applications. For these reasons, a number of other analytical methods have been developed to measure the moisture content of foods that do not rely on direct measurement of the mass of water in a food. Instead, these techniques are based on the fact that the water in a food can be distinguished from the other components in some measurable way.

An appreciation of the principles, advantages and limitations of the various analytical techniques developed to determine the moisture content of foods depends on an understanding of the molecular characteristics of water. A water molecule consists of an oxygen atom covalently bound to two hydrogen atoms (H_2O). Each of the hydrogen atoms has a small positive charge (d+), while the oxygen atom has two lone pairs of electrons that each has a small negative charge (d-). Consequently, water molecules are capable of forming relatively strong hydrogen bonds ($O\text{-}H^{d+}\,\square\,{}^{d-}O$) with four neighboring water molecules. The strength and directionality of these hydrogen bonds are the origin of many of the unique physicochemical properties of water. The development of analytical techniques to determine the moisture content of foods depends on being able to distinguish water (the "analyte") from the other components in the food (the "matrix"). The characteristics of water that are most commonly used to achieve this are: its relatively low boiling point; its high polarity; its ability to undergo unique chemical reactions with certain reagents; its unique electromagnetic absorption spectra; and, its characteristic physical properties (density, compressibility, electrical conductivity and refractive index).

Despite having the same chemical formula (H_2O) the water molecules in a food may be present in a variety of different molecular environments depending on their interaction with the surrounding molecules. The water molecules in these different environments normally have different physiochemical properties:

- *Bulk water*. Bulk water is free from any other constituents, so that each water molecule is surrounded only by other water molecules. It therefore has physicochemical properties that are the same as those of pure water, *e.g.*, melting point, boiling point, density, compressibility, heat of vaporization, electromagnetic absorption spectra.

- *Capillary or trapped water.* Capillary water is held in narrow channels between certain food components because of capillary forces. Trapped water is held within spaces within a food that are surrounded by a physical barrier that prevents the water molecules from easily escaping, *e.g.*, an emulsion droplet or a biological cell. The majority of this type of water is involved in normal water-water bonding and so it has physicochemical properties similar to that of bulk water.
- *Physically bound water.* A significant fraction of the water molecules in many foods are not completely surrounded by other water molecules, but are in molecular contact with other food constituents, *e.g.* proteins, carbohydrates or minerals. The bonds between water molecules and these constituents are often significantly different from normal water-water bonds and so this type of water has different physicochemical properties than bulk water *e.g.*, melting point, boiling point, density, compressibility, heat of vaporization, electromagnetic absorption spectra.
- *Chemically bound water.* Some of the water molecules present in a food may be chemically bonded to other molecules as water of crystallization or as hydrates, *e.g.* $NaSO_4.10H_20$. These bonds are much stronger than the normal water-water bond and therefore chemically bound water has very different physicochemical properties to bulk water, *e.g.*, lower melting point, higher boiling point, higher density, lower compressibility, higher heat of vaporization, different electromagnetic absorption spectra.

Foods are heterogeneous materials that contain different proportions of chemically bound, physically bound, capillary, trapped or bulk water. In addition, foods may contain water that is present in different physical states: gas, liquid or solid. The fact that water molecules can exist in a number of different molecular environments, with different physicochemical properties, can be problematic for the food analyst trying to accurately determine the moisture content of foods. Many analytical procedures developed to measure moisture content are more sensitive to water in certain types of molecular environment than to water in other types of molecular environment.

This means that the measured value of the moisture content of a particular food may depend on the experimental technique used to carry out the measurement. Sometimes food analysts are interested in determining the amounts of water in specific molecular environments (*e.g.*, physically bound water), rather than the total water content. For example, the rate of microbial growth in a food depends on the amount of bulk water present in a food, and not necessarily on the total amount of water present. There are analytical techniques available that can provide some information about the relative fractions of water in different molecular environments (*e.g.*, DSC, NMR, vapor pressure).

Sample Preparation

Selection of a representative sample, and prevention of changes in the properties of the sample prior to analysis, are two major potential sources of error in any food analysis procedure. When determining the moisture content of a food it is important to prevent any loss or gain of water. For this reason, exposure of a sample to the atmosphere, and excessive temperature fluctuations, should be minimized. When samples are stored in containers it is common practice to fill the container to the top to prevent a large headspace, because this reduces changes in the sample due to equilibration with its environment. The most important techniques developed to measure the moisture content of foods are discussed below.

Evaporation Methods

Principles

These methods rely on measuring the mass of water in a known mass of sample. The moisture content is determined by measuring the mass of a food before and after the water is removed by evaporation:

$$\%\text{Moisture} = \frac{M_{\text{INITIAL}} - M_{\text{DRIED}}}{M_{\text{INITIAL}}} \times 100$$

Here, M_{INITIAL} and M_{DRIED} are the mass of the sample before and after drying, respectively. The basic principle of this technique is that water has a lower boiling point than the other major components within foods, *e.g.*,lipids, proteins, carbohydrates and minerals. Sometimes a related parameter, known as the *total solids,* is reported as a measure of the moisture content. The total solids content is a measure of the amount of material remaining after all the water has been evaporated:

$$\%\text{Total Solids} = \frac{M_{\text{DRIED}}}{M_{\text{INITIAL}}} \times 100$$

Thus, %Total solids = (100 - %Moisture). To obtain an accurate measurement of the moisture content or total solids of a food using evaporation methods it is necessary to remove all of the water molecules that were originally present in the food, without changing the mass of the food matrix. This is often extremely difficult to achieve in practice because the high temperatures or long times required to remove all of the water molecules would lead to changes in the mass of the food matrix, *e.g.,* due to volatilization or chemical changes of some components. For this reason, the drying conditions used in evaporation methods are usually standardized in terms of temperature and time so as to obtain results that are as accurate and reproducible as possible given the practical constraints. Using a standard method of sample preparation and analysis helps to minimize sample-to-sample variations within and between laboratories.

Evaporation Devices

The thermal energy used to evaporate the water from a food sample can be provided directly (*e.g.*, transfer of heat from an oven to a food) or indirectly (*e.g.*, conversion of electromagnetic radiation incident upon a food into heat due to absorption of energy by the water molecules).

- *Convection and forced draft ovens.* Weighed samples are placed in an oven for a specified time and temperature (*e.g.* 3 hours at 100 °C) and their dried mass is determined, or they are dried until they reach constant mass. The thermal energy used to evaporate the water is applied directly to the sample *via* the shelf and air that surround it. There are often considerable temperature variations within convection ovens, and so precise measurements are carried out using *forced draft ovens* that circulate the air so as to achieve a more uniform temperature distribution within the oven. Samples that contain significant quantities of carbohydrates that might undergo chemical changes or volatile materials other than water should not be dried in a convection or forced draft oven. Many official methods of analysis are based on forced draft ovens.
- *Vacuum oven.* Weighed samples are placed under reduced pressure (typically 25-100 mm Hg) in a vacuum oven for a specified time and temperature and their dried mass is determined. The thermal energy used to evaporate the water is applied directly to the sample *via* the metallic shelf that it sits upon. There is an air inlet and outlet to carry the moisture lost from the sample out of the vacuum oven, which prevents the accumulation of moisture within the oven. The boiling point of water is reduced when it is placed under vacuum. Drying foods in a vacuum oven therefore has a number of advantages over conventional oven drying techniques. If the sample is heated at the same temperature, drying can be carried out much quicker. Alternatively, lower temperatures can be used to remove the moisture (*e.g.* 70°C instead of 100 °C), and so problems associated with degradation of heat labile substances can be reduced. A number of vacuum oven methods are officially recognized.
- *Microwave oven.* Weighed samples are placed in a microwave oven for a specified time and power-level and their dried mass is weighed. Alternatively, weighed samples may be dried until they reach a constant final mass - analytical microwave ovens containing balances to continuously monitor the weight of a food during drying are commercially available. The water molecules in the food evaporate because they absorb microwave energy, which causes them to become thermally excited. The major advantage of microwave methods over other drying methods is that they are simple to use and rapid to carry out. Nevertheless, care must be taken to standardize the drying procedure and ensure that the microwave energy is applied evenly across the sample. A number of microwave oven drying methods are officially recognized.

- *Infrared lamp drying.* The sample to be analyzed is placed under an infrared lamp and its mass is recorded as a function of time. The water molecules in the food evaporate because they absorb infrared energy, which causes them to become thermally excited. One of the major advantages of infrared drying methods is that moisture contents can be determined rapidly using inexpensive equipment, *e.g.,* 10-25 minutes. This is because the IR energy penetrates into the sample, rather than having to be conducted and convected inwards from the surface of the sample. To obtain reproducible measurements it is important to control the distance between the sample and the IR lamp and the dimensions of the sample. IR drying methods are not officially recognized for moisture content determinations because it is difficult to standardize the procedure. Even so, it is widely used in industry because of its speed and ease of use.

Practical Considerations

- *Sample dimensions.* The rate and extent of moisture removal depends on the size and shape of the sample, and how finely it is ground. The greater the surface area of material exposed to the environment, the faster the rate of moisture removal.
- *Clumping and surface crust formation.* Some samples tend to clump together or form a semi-permeable surface crust during the drying procedure. This can lead to erroneous and irreproducible results because the loss of moisture is restricted by the clumps or crust. For this reason samples are often mixed with dried sand to prevent clumping and surface crust formation.
- *Elevation of boiling point.* Under normal laboratory conditions pure water boils at 100 °C. Nevertheless, if solutes are present in a sample the boiling point of water is elevated. This is because the partial vapor pressure of water is decreased and therefore a higher temperature has to be reached before the vapor pressure of the system equals the atmospheric pressure. Consequently, the rate of moisture loss from the sample is slower than expected. The boiling point of water containing solutes (T_b) is given by the expression, $T_b = T_0 + 0.51m$, where T_0 is the boiling point of pure water and m is the molality of solute in solution (mol/kg of solvent).
- *Water type.* The ease at which water is removed from a food by evaporation depends on its interaction with the other components present. Free water is most easily removed from foods by evaporation, whereas more severe conditions are needed to remove chemically or physically bound water. Nevertheless, these more extreme conditions can cause problems due to degradation of other ingredients which interfere with the analysis (see below).
- *Decomposition of other food components.* If the temperature of drying is too high, or the drying is carried out for too long, there may be decomposition of some of the heat-sensitive components in the food. This will cause a

change in the mass of the food matrix and lead to errors in the moisture content determination. It is therefore normally necessary to use a compromise time and temperature, which are sufficient to remove most of the moisture, but not too long to cause significant thermal decomposition of the food matrix. One example of decomposition that interferes with moisture content determinations is that of carbohydrates.

$$C_6H_{12}O_6 \rightarrow 6C + 6\,H_2O$$

The water that is released by this reaction is not the water we are trying to measure and would lead to an overestimation of the true moisture content. On the other hand, a number of chemical reactions that occur at elevated temperatures lead to water absorption, *e.g.*, sucrose hydrolysis (sucrose + $H_2O \xrightarrow{heat}$ fructose + glucose), and therefore lead to an underestimation of the true moisture content. Foods that are particularly susceptible to thermal decomposition should be analyzed using alternative methods, *e.g.* chemical or physical.

- *Volatilization of other food components.* It is often assumed that the weight loss of a food upon heating is entirely due to evaporation of the water. In practice, foods often contain other volatile constituents that can also be lost during heating, *e.g.*, flavors or odors. For most foods, these volatiles only make up a very small proportion and can therefore be ignored. For foods that do contain significant amounts of volatile components (*e.g.* spices and herbs) it is necessary to use alternative methods to determine their moisture content, *e.g.*, distillation, chemical or physical methods.
- *High moisture samples.* Food samples that have high moisture contents are usually dried in two stages to prevent "spattering" of the sample, and accumulation of moisture in the oven. Spattering is the process whereby some of the water jumps out of the food sample during drying, carrying other food constituents with it. For example, most of the moisture in milk is removed by heating on a steam bath prior to completing the drying in an oven.
- *Temperature and power level variations.* Most evaporation methods stipulate a definite temperature or power level to dry the sample so as to standardize the procedure and obtain reproducible results. In practice, there are often significant variations in temperatures or power levels within an evaporation instrument, and so the efficiency of the drying procedure depends on the precise location of the sample within the instrument. It is therefore important to carefully design and operate analytical instruments so as to minimize these temperature or power level variations.
- *Sample pans.* It is important to use appropriate pans to contain samples, and to handle them correctly, when carrying out a moisture content analysis. Typically aluminum pans are used because they are relatively cheap and have a high thermal conductivity. These pans usually have lids to prevent spattering of the sample, which would lead to weight loss and

therefore erroneous results. Pans should be handled with tongs because fingerprints can contribute to the mass of a sample. Pans should be dried in an oven and stored in a descicator prior to use to ensure that no residual moisture is attached to them.

Advantages and Disadvantages

- *Advantages:* Precise; Relatively cheap; Easy to use; Officially sanctioned for many applications; Many samples can be analyzed simultaneously
- *Disadvantages:* Destructive; Unsuitable for some types of food; Time consuming

Distillation Methods

Principles

Distillation methods are based on *direct* measurement of the amount of water removed from a food sample by evaporation: %Moisture = 100 ($M_{WATER}/M_{INITIAL}$). In contrast, evaporation methods are based on*indirect* measurement of the amount of water removed from a food sample by evaporation: %Moisture = 100 ($M_{INITIAL} - M_{DRIED})/M_{INITIAL}$. Basically, distillation methods involve heating a weighed food sample ($M_{INITIAL}$) in the presence of an organic solvent that is immiscible with water. The water in the sample evaporates and is collected in a graduated glass tube where its mass is determined (M_{WATER}).

Dean and Stark Method

Distillation methods are best illustrated by examining a specific example: the Dean and Stark method. A known weight of food is placed in a flask with an organic solvent such as xylene or toluene. The organic solvent must be insoluble with water; have a higher boiling point than water; be less dense than water; and be safe to use. The flask containing the sample and the organic solvent is attached to a condenser by a side arm and the mixture is heated. The water in the sample evaporates and moves up into the condenser where it is cooled and converted back into liquid water, which then trickles into the graduated tube. When no more water is collected in the graduated tube, distillation is stopped and the volume of water is read from the tube.

Practical Considerations

There are a number of practical factors that can lead to erroneous results: (i) emulsions can sometimes form between the water and the solvent which are difficult to separate; (ii) water droplets can adhere to the inside of the glassware, (iii) decomposition of thermally labile samples can occur at the elevated temperatures used.

Advantages and Disadvantages

- *Advantages:* Suitable for application to foods with low moisture contents; Suitable for application to foods containing volatile oils, such as herbs or spices, since the oils remain dissolved in the organic solvent, and therefore do not interfere with the measurement of the water; Equipment is relatively cheap, easy to setup and operate; Distillation methods have been officially sanctioned for a number of food applications.
- *Disadvantages:* Destructive; Relatively time-consuming; Involves the use of flammable solvents; Not applicable to some types of foods.

Chemical Reaction Methods

Reactions between water and certain chemical reagents can be used as a basis for determining the concentration of moisture in foods. In these methods a chemical reagent is added to the food that reacts specifically with water to produce a measurable change in the properties of the system, *e.g.*, mass, volume, pressure, pH, color, conductivity. Measurable changes in the system are correlated to the moisture content using calibration curves. To make accurate measurements it is important that the chemical reagent reacts with all of the water molecules present, but not with any of the other components in the food matrix. Two methods that are commonly used in the food industry are the *Karl-Fisher titration* and *gas production* methods. Chemical reaction methods do not usually involve the application of heat and so they are suitable for foods that contain thermally labile substances that would change the mass of the food matrix on heating (*e.g.*, food containing high sugar concentrations) or foods that contain volatile components that might be lost by heating (*e.g.* spices and herbs).

Karl-Fisher Method

The Karl-Fisher titration is often used for determining the moisture content of foods that have low water contents (*e.g.* dried fruits and vegetables, confectionary, coffee, oils and fats). It is based on the following reaction:

$$2H_2O + SO_2 + I_2 \rightarrow H_2SO_4 + 2HI$$

This reaction was originally used because HI is colorless, whereas I_2 is a dark reddish brown color, hence there is a measurable change in color when water reacts with the added chemical reagents. Sulfur dioxide and iodine are gaseous and would normally be lost from solution. For this reason, the above reaction has been modified by adding solvents (*e.g.*, C_5H_5N) that keep the S_2O and I_2 in solution, although the basic principles of the method are the same. The food to be analyzed is placed in a beaker containing solvent and is then titrated with *Karl Fisher reagent* (a solution that contains iodine). While any water remains in the sample the iodine reacts with it and the solution remains colorless (HI), but once all the water has been used up any additional iodine is observed as a dark red brown color (I_2). The volume of iodine solution required to titrate the water is measured and can be related to the moisture content using a pre-prepared calibration curve. The precision of the technique can be improved by using electrical methods to follow the end-point of the reaction, rather than observing a color change. Relatively inexpensive commercial instruments have been developed which are based on the Karl-Fisher titration, and some of these are fully automated to make them less labor intensive.

Gas Production Methods

Commercial instruments are also available that utilize specific reactions between chemical reagents and water that lead to the production of a gas. For example, when a food sample is mixed with powdered calcium carbide the amount of acetylene gas produced is related to the moisture content.

$$CaC_2 + 2H_2O \rightarrow C_2H_2(gas) + Ca(OH)_2$$

The amount of gas produced can be measured in a number of different ways, including (i) the volume of gas produced, (ii) the decrease in the mass of the sample after the gas is released, and (iii) the increase in pressure of a closed vessel containing the reactants.

Physical Methods

A number of analytical methods have been developed to determine the moisture content of foods that are based on the fact that water has appreciably different bulk physical characteristics than the food matrix, *e.g.*density, electrical conductivity or refractive index. These methods are usually only suitable for analysis of foods in which the composition of the food matrix does not change significantly, but the ratio of water-to-food matrix changes. For example, the water content of oil-in-water emulsions can be determined by measuring their density or electrical conductivity because the density and electrical conductivity of water are significantly higher than those of oil. If the composition of the food matrix changes as well as the water content, then it may not be possible to accurately determine the moisture content of the food because more than one food composition may give the same value for the physical property being measured. In these cases, it may be possible to use a combination of two or more physical methods to determine the composition of the food, *e.g.*, density measurements in combination with electrical conductivity measurements.

Spectroscopic Methods

Spectroscopic methods utilize the interaction of electromagnetic radiation with materials to obtain information about their composition, *e.g.*, X-rays, UV-visible, NMR, microwaves and IR. The spectroscopic methods developed to measure the moisture content of foods are based on the fact that water absorbs electromagnetic radiation at characteristic wavelengths that are different from the other components in the food matrix. The most widely used physical methods are based on measurements of the absorption of microwave or infrared energy by foods. Microwave and infrared radiation are absorbed by materials due to their ability to promote the vibration and/or rotation of molecules. The analysis is carried out at a wavelength where the water molecules absorb radiation, but none of the other components in the food matrix do. A measurement of the absorption of radiation at this wavelength can then be used to determine the moisture content: the higher the moisture content, the greater the absorption. Instruments based on this principle are commercially available and can be used to determine the moisture content in a few minutes or less. It is important not to confuse infrared and microwave *absorption* methods with infrared lamp and microwave *evaporation* methods. The former use low energy waves that cause no physical or chemical changes in the food, whereas the latter use high-energy waves to evaporate the water. The major advantage of these methods is that they are capable of rapidly determining the moisture content of a food with little or no sample preparation and are therefore particularly useful for quality control purposes or rapid measurements of many samples.

Methods to Determine Water in Different Molecular Environments

The overall water content of a food is sometimes not a very reliable indication of the quality of a food because the water molecules may exist in different environments within foods, *e.g.*, "bound" or "free". Here "bound water" refers to water that is physically or chemically bound to other food components, whereas "free water" refers to bulk, capillary or entrapped water. For example, the microbial stability or physicochemical properties of a food are often determined by the amount of free water present, rather than by the total amount of water present. For this reason, it is often useful for food scientists to be able to determine the amount of water in different molecular environments within a food. A variety of analytical methods are available that can provide this type of information.

Vapor Pressure Methods

A physical parameter that is closely related to the amount of free water present in a food is the water activity:

$$a_w = \frac{P}{P_0}$$

where, P is the partial pressure of the water above the food and P_0 is the vapor pressure of pure water at the same temperature. Bound water is much less volatile than free water, and therefore the water activity gives a good indication of the amount of free water present. A variety of methods are available for measuring the water activity of a sample based on its vapor pressure. Usually, the sample to be analyzed is placed in a closed container and allowed to come into equilibrium with its environment. The water content in the headspace above the sample is then measured and compared to that of pure water under the same conditions.

Thermogravimetric Methods

Thermogravimetric techniques can be used to continuously measure the mass of a sample as it is heated at a controlled rate. The temperature at which water evaporates depends on its molecular environment: free water normally evaporates at a lower temperature than bound water. Thus by measuring the change in the mass of a sample as it loses water during heating it is often possible to obtain an indication of the amounts of water present in different molecular environments.

Calorimetric Methods

Calorimetric techniques such as differential scanning calorimetry (DSC) and differential thermal analysis (DTA) can be used to measure changes in the heat absorbed or released by a material as its temperature is varied at a controlled rate. The melting point of water depends on its molecular environment: free water normally melts at a higher temperature than bound water. Thus by measuring the enthalpy change of a sample with temperature it is possible to obtain an indication of the amounts of water present in different molecular environments.

Spectroscopic Methods

The electromagnetic spectrum of water molecules often depends on their molecular environment, and so some spectroscopy techniques can be used to measure the amounts of water in different environments. One of the most widely used of these techniques is nuclear magnetic resonance (NMR). NMR can distinguish molecules within materials based on their molecular mobility, *i.e.,* the distance they move in a given time. The molecular mobility of free water is appreciably higher than that of bound water and so NMR can be used to provide an indication of the concentrations of water in "free" and "bound" states.

Chapter 8

Determination of Vitamin C

Ascorbic acid, or vitamin C, is very important for our bodies. Vitamin C is required by our bodies to produce collagen in our connective tissues. Vitamin C thus helps us to maintain healthy teeth, blood vessels, and bones. Vitamin C has also been found to function as an antioxidant, and it aids in the prevention of cancer.

Antioxidants are molecules that can protect our cells from being damaged. Antioxidants such as vitamin C are able to absorb harmful reactive molecules before they can damage our tissues. In this investigation, a titration will be used to determine the amount of vitamin C that is contained in "vitamin C" tablets and in different foods and juices. The recommended daily amount of vitamin C is 60 mg.

Materials

- • 250 mL beaker or Erlenmeyer flask
- • Vitamin C solution (2 mg/mL), obtained by grinding one 500 mg tablet and dissolving in 250 mL of distilled water
- Dropper bottle
- 10 mL graduated cylinder
- 10 g of starch
- Distilled water bottle
- 20 mL of 0.010 M KIO3 solution
- 1.0 g KI
- 5.0 mL 1.0 M HCl

- Orange juice and grapefruit juice
- Citrus fruits and juicer
- Cheesecloth
- Medium funnel

We will determine the amount of vitamin C using a simple titration method. The chemical name for vitamin C is ascorbic acid. This procedure relies on a series of chemical reactions that cause an observable color change to occur. To help you better understand what you are seeing when the color change occurs, it is helpful to review some relevant chemical reactions.

In equation two, Vitamin C reacts with iodine (I2) and becomes oxidized. This is the most important chemical reaction taking place. However, to produce the iodine (I2) we must first carry out another reaction. Adding solid potassium iodide (KI) and hydrochloric acid (HCl) and the titrant, potassium iodate (KIO3), creates iodine and water according to the following reaction:

$IO3- + 5I- + 6H+ . 3I2 + 3H2O$

As soon as the iodine (I2) is created it reacts with the vitamin C, causing the vitamin C to become oxidized.

After all of the vitamin C has reacted, the I2 concentration builds up and then begins to react with I– in solution to form the triiodide ion (I3–). The triiodide ion complexes with the starch to give a deep bluish-black color that we can see:

$I2 + I- \rightarrow I3 -$ and $I3 - +$ starch $\rightarrow$ iodide–starch complex deep blue color

When this color appears, we know all of the vitamin C has reacted. It is the formation of the triiodide ion that signals the endpoint of the titration.

Once we know how much titrant (KIO3) is required to reach the endpoint, we can use the volume of KIO3 to determine the moles of KIO3. Using stoichiometry we can use the moles of KIO3 to find the moles of iodine (I2) and therefore the moles of vitamin C present in our experiment.

Part 1: Calibrate your dropper bottle or pipette

1. We need to accurately determine the number of drops in 1.0 mL. To accomplish this, use either a dropper bottle or a plastic beral pipet. [In Part 2, be sure you add the titrant (KIO3) using the same dropper and method].
2. To calibrate your dropper, slowly add drops of water to a small 10.0 mL graduated cylinder. Stop adding drops when the meniscus is on the 1.0 mL mark. Record the number of drops you added.
3. Next count how many drops it takes to get to the 2.0 mL mark. Record your measurement. If the number of drops is in close agreement, take the average. If your number of drops does not agree, repeat Steps 2 and 3.

Part 2: Verifying the amount of vitamin C in a commercial tablet

1. 1. Measure 25.0 mL of the vitamin C stock solution, and add it to a 250 mL Erlenmeyer flask. Your instructor will have prepared this ahead of time.

Record the concentration of the stock solution in your lab notebook.

2. Prepare a 1% starch solution.
3. Measure 1.0 g of KI solid and add it to the Erlenmeyer flask. Swirl to dissolve.
4. Add 5.0 mL of 1.0 M HCl to the Erlenmeyer flask.
5. Add 2–3 mL of the fresh 1.0% starch solution indicator to the Erlenmeyer flask. Swirl to mix.
6. Obtain a dropper bottle of 0.010 M KIO3 titrant solution. This dropper bottle should be the same type as the one you used in Part 1. If your instructor has the titrant in a stock bottle, measure about 25.0 mL of the 0.010 M KIO3 titrant solution in a clean, dry beaker and add it to a dropper bottle yourself.
7. Now you are ready to titrate. Add 0.010 M KIO3 solution drop by drop to the Erlenmeyer flask. Swirl after every few drops. Initially you will see the deep blue color appear in the area that the drop is added. Once you begin to see a deep blue color that persists throughout the solution, add the titrant more slowly and swirl after every drop. Continue adding titrant until you obtain a permanent blue color that lasts about 60 s. Record the total number of drops you added. The solution may lighten after the 60 s, but it should remain fairly deep blue. Determining the "right" color takes practice and is somewhat objective. Do your best, and then repeat the experiment. After two or three trials, you will feel more confident at determining your endpoint.

Part 3: How much vitamin C is there in different fruits and juices?

Here we will determine the vitamin C content of a food of your choice. You may use citrus fruits, fresh squeezed or prepared, apple juice, or any drink containing ascorbic acid. Real lemon juice also makes a nice sample to test, and the amount of vitamin C is on the label.

1. To begin you may use fresh or prepared juice. You will need about 25 mL of juice to have enough vitamin C for a good measurement. If you use freshly squeezed juice, it is helpful to strain out the pulp with cheesecloth. An easy way to strain out the pulp is to use a funnel lined with two or three layers of cheesecloth. Pour the sample of juice through the cheesecloth and collect the juice. Record the exact volume of juice you will titrate.
2. Add your juice to a 250 mL Erlenmeyer flask.
3. Titrate your sample using the same steps you used to titrate the vitamin C tablet. Be sure to add 1.0 g of solid KI, 5.0 mL of 1.0 M HCl and 2–3 mL of starch solution. Note: You may not see the deep blue color because of the color of your fruit juice. For instance, it may appear brownish instead of blue. This is fine; just be sure you look for the same color change in each trial. That is the key.

4. Repeat your experiment two or three times to obtain an accurate sense of the reproducibility of your results. Record your results and find the average amount of vitamin C in your sample.

Part 4: Thinking about it

1. Make a data table that records the number of drops of 0.010 M KIO3 needed to titrate 25 mL of the vitamin C stock solution.
2. Convert your drops to milliliters using the conversion you determined in Part 1. Record this in your data table.
3. Does it matter how much distilled water was added to your sample? Explain.
4. Which contained more vitamin C: the tablet or the food? How do you know?
5. Look carefully at the first equation on page 113. Copy this equation into your lab notebook. Which substance was the reducing agent? Explain. Which substance was the oxidizing agent? Explain.

Part 5: What did you obtain for results?

1. Use the volume of your KIO3 titrant added in liters and the known molarity (0.010 M) to calculate the moles of KIO3 required to titrate the vitamin C sample, as shown in this example:
2. Using the second equation on page 113 and stoichiometry, convert the moles of KIO3 to moles of I2 produced. See the example calculation shown in Step 3.
3. Calculate your moles of vitamin C, from the amount of I2 produced in Step 2.
4. Calculate the grams of vitamin C from the moles of vitamin C you determined.
5. Convert your grams to milligrams by dividing by 1,000. This number tells you how many milligrams of vitamin C were in your 25.0 mL sample of the vitamin C stock solution, or your food.
6. Which food contained the highest amount of vitamin C? Explain.
7. The stock solution contained a 500 mg tablet of vitamin C and 250 mL of distilled water. Combine your results with those of your classmates and see how close your results came to this value. For example, if each group titrated 25.0 mL of stock solution, then the total milligrams in 10 trials would give you a total volume of 250 mL of solution, and this should give you a value close to 500 mg of vitamin C. To estimate, you could also take your trial average and multiply the number of milligrams by 10 to see if you are close to 500 mg.
8. Based on your result, does a 500 mg tablet really contain 500 mg of vitamin C?
9. 9. If the recommended daily allowance of vitamin C is 60 mg, how much of your food would you need to consume to get your daily allowance? Show your calculation.

Chapter 9

Determination of Sodium and Potassium

The SCAN-test Method replaces SCAN-CM 63:00/P 83:00 from which it differs in that a number of withdrawn SCAN-test Methods, referred to in this method, have been replaced by the corresponding ISO standards. In SCAN-C 6, the ignition temperature was 575 °C but in the replacing ISO 1762 the ignition temperature is 525 °C so for that reason the temperature has been changed.

The SCAN-test Method describes a procedure for the determination of the contents of acid-soluble sodium and potassium using Atomic Absorption Spectroscopy (AAS) using a flame. Two alternative digestion techniques are specified: - ignition at 525 °C and - wet digestion in an autoclave or microwave oven.

Note – The determination can also be made by inductively coupled plasma spectroscopy (ICP-AES) or flame atomic emission spectroscopy (AES).

1. Scope

The Method is applicable to most types of pulp, paper and board having sodium and potassium contents that are acid-soluble by using one of the wet digestion procedures described in this Method.

The method is not suitable for samples containing kaolin as coating or filler pigment because of the difficulty in dissolving sodium and potassium quantitatively.

Wet digestion in an open system is not included in this method. The technique is however applicable to most types of samples. It is not recommended for potassium determination in samples containing clay because it gives too low results.

2. References

ISO 638 Pulps – Determination of dry matter content ISO 1762 Paper, board and pulps – Determination of residue (ash) on ignition at 525 °C

Note – SCAN-test has withdrawn a number of test methods and refers instead to the corresponding ISO and/or EN Standards.

3. Definitions

For the purpose of this Method, the following definitions apply:

3.1 Sodium content, acid-soluble – The sodium content in the pulp, paper or board obtained after either:

- Ignition the material at 525 °C and treating the residue with 6 mol/l hydrochloric acid as specified in this method, or
- Wet digestion of the material with concentrated nitric acid in a closed vessel as specified in this method.

3.2 Potassium content, acid-soluble – The potassium content in the pulp, paper or board obtained after either:

- Ignition the material at 525 °C and treating the residue with 6 mol/l hydrochloric acid as specified in this method, or

- Wet digestion of the material with concentrated nitric acid in a closed vessel as specified in this method.

4. Principle

4.1 Alternative A – ignition

The material is ignited at (525 ± 25) ºC. The residue is treated with 6 mol/l hydrochloric acid (HCl) and the sodium and potassium contents in the solution are determined by Atomic Absorption Spectroscopy (AAS) using a flame.

4.2 Alternative B – wet digestion (closed system)

The material is treated with concentrated nitric acid (HNO3) in a closed vessel at elevated temperature and pressure. The resulting solution is diluted and the sodium and potassium contents are determined by Atomic Absorption Spectroscopy (AAS) using a flame.

The method specifies two alternative wet-digestion techniques: - digestion in an autoclave; - digestion in a laboratory microwave oven.

5. Reagents and Chemicals

Use only chemicals of analytical grade.

5.1 Water, of high purity, distilled or deionized.

5.2 Hydrochloric acid, c(HCl) = 6 mol/l. Dilute 500 ml of concentrated hydrochloric acid, (37 %, density 1,19 g/ml) in 500 ml of water (5.1).

5.3 Concentrated nitric acid, (HNO_3), 65 %, density 1,40 g/ml.

5.4 Sodium stock solution, 1000 mg/l. Ignite a portion of anhydrous sodium sulphate, Na_2SO_4, at 550 °C, in a crucible of platinum or porcelain. Allow to cool to room temperature in a desiccator.

Weigh (3,089 ± 0,005) g, using an analytical balance, and dissolve in water (5.1). Dilute to 1 litre in a volumetric flask. Store in a polyethylene bottle.

5.4.1 Sodium standard solution, 50 mg/l. Dilute 5,00 ml of the sodium stock solution (5.4) to 100 ml in a volumetric flask.

5.5 Potassium stock solution, 1000 mg/l. Ignite a portion of anhydrous potassium sulphate, K_2SO_4, at 550 °C, in a crucible of platinum or porcelain. Allow to cool to room temperature in a desiccator.

Weigh (2,228 ± 0,005) g, using an analytical balance, and dissolve in water (5.1). Dilute to 1 litre in a volumetric flask. Store in a polyethylene bottle.

5.5.1 Potassium standard solution, 50 mg/l. Dilute 5,00 ml of the potassium stock solution (5.5) to 100 ml in a volumetric flask.

5.6 Cesium solution, c(Cs) = 50 g/l. In a 1000 ml volumetric flask, dissolve 63,5 g of cesium chloride, CsCl, in water (5.1) and fill up to the mark with water.

Note – Commercially available solutions may be used.

6. Apparatus

6.1 Alternative A and B

6.1.1 Ordinary laboratory equipment. Laboratory vessels made of plastics are recommended as well as ash-free and acid-washed filter paper. A suitable filter paper is Munktell No 00A but other similar qualities may be used.

Note 1 – Solutions can be diluted in acid-washed glassware bottles, but solutions should be stored in plastic bottles.

6.1.2 Atomic absorption spectrometer, with a burner for air-acetylene and with hollow-cathode lamps for the elements to be determined (multi-element lamps may be used).

6.2 Alternative A – ignition

6.2.1 Furnace, capable of maintaining a temperature of (525 ± 25) °C. It is recommended the furnace be fitted with a hood or other means of evacuating smoke and fumes.

6.2.2 Dishes of platinum or quartz. Wash the dishes thoroughly. Remove any spots in the platinum dishes by cleaning with fine sand. Boil the dishes with hydrochloric acid (5.2) and avoid all contact with metals other than platinum.

6.3 Alternative B – wet digestion

Apparatus for wet digestion, either:

6.3.1 Autoclave, with an inner vessel of PTFE (polytetrafluoroethylene) provided with a lid of PTFE, and a heating block that can maintain the autoclave at a temperature of (160 ± 5) °C. The block must be provided with safety arrangements that prevent it from being overheated.

Note 2 - The use of an oven instead of the heating block is not recommended because of the hazard involved in removing the hot autoclaves from the oven at the end of the heating period.

or

6.3.2 Laboratory microwave oven equipped with a rotating vessel holder and with programming facilities. The digestion vessels, made of PTFE, shall have a volume of at least 80 ml and either low-pressure or high-pressure vessels may be used. Follow the safety regulations given by the oven supplier. Since the test portion size influences the change of pressure and temperature during the wet digestion, the oven should preferably be equipped with temperature- and pressure-control facilities. The oven shall be able to keep the temperature constant at 175 °C.

7. Sample Preparation

The sampling procedure is not covered by this Method. Make sure that the test portions taken are representative of the sample received.

Carry out the determination in duplicate. Tear the air-dry sample into pieces of suitable size. Weigh a separate test portion for the determination of dry matter content in accordance with ISO 638.

8. Ignition and Wet Digestion

8.1 Alternative A – ignition

8.1.1 Weighing. In a clean dish (6.2.2), weigh to the nearest 0,01 g a test portion of about 2 g to 5 g.

8.1.2 Ignition. Use a hot plate, a temperature-controlled oven or a low flame of a gas burner and char the test portion until it is well carbonised. The temperature shall be (525 ± 25) °C. If the dish is too small to hold the entire test portion, gently burn it off in smaller portions. Take care not to let any ash escape from the dish. Place the dish in a furnace and ignite it at a temperature of (525 ± 25) °C for a period of 3 h, or longer if needed to burn away all the carbon.

Note 1 - In order to avoid flames, the dish can be covered by a lid. The lid must not be so tight that reducing conditions occur in the dish, since acid-insoluble compounds may then be formed.

8.1.3 Dissolving the ignited test portion in acid. To avoid splattering, cautiously moisten the ash with water (5.1) and add 5 ml of hydrochloric acid (5.2) to the dish.

Evaporate to dryness on a boiling water-bath or equivalent device. Repeat this procedure once.

Add 2,5 ml of the acid (5.2) to the dry ignited test portion and heat the dish, covered with a watch glass to prevent evaporation.

Filter the contents of the dish through a filter paper and collect the filtrate in a 25 ml volumetric flask. To ensure that the transfer is quantitative, add another portion of 2,5 ml of acid (5.2) to the dish and heat it again. Filter this last portion of acid to the main portion in the volumetric flask with the aid of some water (5.1).

Wash the insoluble residue on the filter paper with water (5.1) and collect the filtrate in the volumetric flask. Fill to the mark and mix. The acid concentration in the solution is about 1,2 mol/l.

8.1.4 Blank. Run a blank with the same chemicals and the same procedure as in 8.1.3 but without any test portion.

8.2 Alternative B – wet digestion (closed system)

8.2.1 Weighing. Weigh to the nearest 1 mg a test portion of about 0,5 g to 2 g, or the maximum amount permissible within safety limits for the equipment used.

8.2.2 Wet digestion. Transfer the test portion to the PTFE vessel. Add 10 ml concentrated nitric acid (5.3) and wait 5 min. Close the vessel with its lid and place it in the autoclave or in the laboratory microwave oven as relevant.

Warning – Esters and alcohols can form explosive compounds with nitric acid. If the sample contains considerable amounts of such compounds, digestion with nitric acid should not be undertaken.

If an autoclave is used: Close the autoclave as instructed by the manufacturer and place it in the heating block. Heat it at a temperature of (160 ± 5) °C for 2 h. Allow the heating block and the autoclave to cool and, with caution, open the autoclave under a hood.

Follow the safety regulations for the equipment.

If a laboratory microwave oven is used: The exact power and time required may vary with the equipment used and the number of samples taken. Heat the closed digestion vessel with its contents, step by step and hold the temperature at (175 ± 5) °C for at least 5 min. It is essential that the organic content of the test portion is completely digested. Allow the vessel to cool and, with caution, open it under a hood.

Note 2 - Check the vessel and the sample after the wet digestion to ensure that no leakage has occurred.

Allow nitrous fumes to escape from the wet digestion vessel. If the solution contains any suspended matter remove it by filtration or centrifugation. Transfer the clear solution quantitatively into a 25 ml volumetric flask and fill to the mark with water (5.1) and mix by shaking. The acid concentration in the solution is about 3 mol/l. (As a rough estimate, half of the added acid is consumed in the oxidation of the organic matter).

8.2.3 Blank. Run a blank with the same chemicals as those added to the wet-digestion but without any test portion.

9. Atomic Absorption Procedure

9.1 Preparation of calibration solutions

It is important that the acid and the cesium concentrations are the same in the calibration and in the test solutions after the final dilution step, since the acid concentration influences the signal. This is particularly important when potassium is to be determined.

Use the same type of acid in the calibration and sample solutions, e.g. - Hydrochloric acid after ignition; - Nitric acid after wet digestion.

Prepare a series of at least 4 calibration solutions by diluting v ml of sodium standard solution (5.4.1) or of potassium standard solution (5.5.1) to 100 ml with water

(5.1) in volumetric flasks. Before filling up to the mark, add 2 ml cesium solution (5.6) and either 10 ml hydrochloric acid (5.2) or 4 ml nitric acid (5.3). Select the volumes v so that the working range of the atomic absorption spectrometer is covered. Prepare also a zero solution (v = 0), similar to the calibration solutions, but containing no added elements. It should not be confused with the blank. An example of suitable calibration solutions is given in Table 1.

Note - A combined calibration solution with both sodium and potassium may be used.

Suitable calibration solutions. By using the procedure described above and the recommendations given in this table, the cesium and acid concentrations in the solutions become 1 g/l and about 0,6 mol/l respectively.

Calibration solutions zero No 1 No 2 No 3 No 4

Conc. of Na or K, mg/l Standard solution, ml (5.4.1 or 5.5.1) 0 0,5 1,0 1,5 2,0 0 1,0 2,0 3,0 4,0

9.2 Determination of sodium and potassium

Dilute the sample solution until the sodium and potassium concentrations are within the range covered by the calibration solutions. In the final dilution stage, adjust the cesium concentration in the sample to 1 g/l by addition of an appropriate amount of cesium solution (5.6). Adjust the acid concentration in the sample to about 0,6 mol/l by addition of appropriate amounts of hydrochloric acid (5.2) or nitric acid (5.3). Notice that the acid concentration in the undiluted samples is about 1,2 mol/l after ignition (8.1.3) and 3 mol/l after wet digestion (8.2.2). This means that the sample solutions should be diluted at least twice and five times respectively to obtain the same acid concentration as in the calibration solutions.

Carry out the spectrometric measurements of the calibration solutions, of the sample solutions and of the blank solution. Adjust the instrument reading to zero by using the zero solution. Following the instructions of the spectrometer, measure the absorbance of sodium and potassium at 589,0 nm and 766,5 nm respectively in an air-acetylene flame.

Most instruments have a system for automatic evaluation of results. If the instrument has no such system, plot the absorbance of the calibration solutions against the elemental content in these solutions. Correct the absorbance of sodium and potassium in the sample solutions for the blank. Note the concentration of sodium and potassium in the sample solutions, obtained from the calibration graph, as c mg/l.

10. Calculation

Calculate the acid-soluble sodium or potassium content from the expression:

$c \cdot f \cdot V\ X = [1]\ m$

where X is the sodium or potassium content of the pulp, paper or board, in milligrams per kilogram;

c is the concentration of sodium or potassium in the sample solution, corrected for the blank, as obtained from the calibration graph, in milligrams per litre; f is the dilution factor (flask volume/aliquot);

V is the volume of the sample solution in millilitres (standard 25 ml);

m is the amount of sample taken, oven dry basis, in grams.

Calculate the mean of the parallel determinations for each element. Report the content with two significant figures for values above 1 mg/kg and with one significant figure in the range 0,1 mg/kg to 1 mg/kg.

11. Report

The test report shall include reference to this SCAN-test Method and the following particulars:

(a) date and place of testing;

(b) identification mark of the sample tested;

(c) the result;

(d) any departure from the standard procedure and

any other circumstances that may have affected

the results.

12. Precision

The precision data given include results from the determination of the acid-soluble sodium and potassium content by both alternative digestion techniques, ignition and wet digestion.

12.1 Repeatability. Five samples were analysed six times in the same laboratory. The following results, mean value and coefficient of variation (CV), were obtained:

Repeatability for the determination of sodium

Sample type Ignition Wet digestion

mean, CV, mean, CV,

mg/kg % mg/kg %

Unbleached pulp 1860 1,0 1840 1,3

Bleached pulp 720 2,0 700 1,4

Kraft liner with recycled pulp 1080 1,0 1060 1,0

Copy paper with calcium carbonate 920 0,9 940 1,3

Paper with kaolin 1290 2,6 1390 2,1

Repeatability for the determination of potassium

Sample type Ignition Wet digestion

mean, CV, mean, CV,

mg/kg % mg/kg %

Unbleached pulp 26 1,6 25 2,6

Bleached pulp 4,4 9,3 5,1 26

Kraft liner with recycled pulp 106 3,2 129 3,2

Copy paper with calcium carbonate 40 3,8 44 2,9

Paper with kaolin 960 1,3 1140* 3,5

12.2 Reproducibility. Five samples were analysed at different laboratories. Results obtained after wet digestion include both digestion in a closed system and in some cases even digestion in an open system. The results, mean value and coefficient of variation (CV) between laboratories, were as follows (the figures in parentheses give the number of laboratories using each method):

Reproducibility for the determination of sodium

Sample type Ignition Wet digestion

mean, CV, mean, CV,

mg/kg % mg/kg %

Unbleached pulp 1990 (6) 6,3 1970 (5) 5,1

Bleached pulp 740 (6) 2,9 730 (5) 3,3

Kraft liner with 1110 (7) 3,7 1130 (7) 4,6

recycled pulp

Copy paper with 1000 (7) 6,0 970 (7) 4,5

calcium carbonate

Paper with kaolin 1310 (7) 18 1360 (7) 19

Reproducibility for the determination of potassium

Sample type Ignition Wet digestion

mean, CV, mean, CV,

mg/kg % mg/kg %

Unbleached pulp 26 (6) 11 27 (5) 4,9

Bleached pulp 4,8 (6) 22 5,0 (5) 9,1

Kraft liner with recycled pulp 110 (8) 11 114 (10) 13

Copy paper with calcium carbonate 44 (9) 18 41 (10) 7,8

Paper with kaolin 910 (7) 28 1260*(5) 15

Chapter 10

Determination of Flavour with Nose

Flavour is the sensory impression of a food or other substance, and is determined mainly by the chemical senses of taste and smell. The "trigeminal senses", which detect chemicalirritants in the mouth and throat as well as temperature and texture, are also very important to the overall Gestalt of flavor perception. The flavor of the food, as such, can be altered with natural or artificial flavorants, which affect these senses.

Flavorant is defined as a substance that gives another substance flavor, altering the characteristics of the solute, causing it to become sweet, sour, tangy, etc.

Of the three chemical senses, smell is the main determinant of a food item's flavor. While the taste of food is limited to sweet, sour, bitter, salty, umami (savory) pungent or piquant, and metallic – the seven basic tastes – the smells of a food are potentially limitless. A food's flavor, therefore, can be easily altered by changing its smell while keeping its taste similar. Nowhere is this better exemplified than in artificially flavored jellies, soft drinks and candies, which, while made of bases with a similar taste, have dramatically different flavors due to the use of different scents or fragrances. The flavorings of commercially produced food products are typically created by flavorists.

Although the terms *"flavoring"* or *"flavorant"* in common language denote the combined chemical sensations of taste and smell, the same terms are usually used in the fragrance and flavors industry to refer to edible chemicals and extracts that alter the flavor of food and food products through the sense of smell. Due to the high cost or unavailability of natural flavor extracts, most commercial flavorants are *nature-identical,* which means that they are the chemical equivalent of natural flavors but chemically synthesized rather than being extracted from the source materials. Identification of nature-identical flavorants are done using technology such as headspace techniques.

Flavorants or Flavorings

Flavorings are focused on altering the flavors of natural food product such as meats and vegetables, or creating flavor for food products that do not have the desired flavors such as candies and other snacks. Most types of flavorings are focused on scent and taste. Few commercial products exist to stimulate the trigeminal senses, since these are sharp, astringent, and typically unpleasant flavors.

There are three principal types of flavorings used in foods, under definitions agreed in the E.U. and Australia.

Type	*Description*
Natural flavoring substances	Flavoring substances obtained from plant or animal raw materials, by physical, microbiological or enzymatic processes. They can be either used in their natural state or processed for human consumption, but cannot contain any nature-identical or artificial flavoring substances.
Nature-identical flavoring substances	Flavoring substances that are obtained by synthesis or isolated through chemical processes, which are chemically and organoleptically identical to flavoring substances naturally present in products intended for human consumption. They cannot contain any artificial flavoring substances.
Artificial flavoring substances	Flavoring substances not identified in a natural product intended for human consumption, whether or not the product is processed. These are typically produced by fractional distillation and additional chemical manipulation of naturally sourced chemicals, crude oil or coal tar. Although they are chemically different, in sensory characteristics are the same as natural ones.

Regulations on Natural Flavoring

UK Food Law defines a natural flavor as:

a flavouring substance (or flavouring substances) which is (or are) obtained, by physical, enzymatic or microbiological processes, from material of vegetable or animal origin which material is either raw or has been subjected to a process normally used in preparing food for human consumption and to no process other than one normally so used.

The U.S. Code of Federal Regulations describes a "natural flavorant" as: the essential oil, oleoresin, essence or extractive, protein hydrolysate, distillate, or any product of roasting, heating or enzymolysis, which contains the flavoring constituents derived from a spice, fruit or fruit juice, vegetable or vegetable juice, edible yeast, herb, bark, bud, root, leaf or any other edible portions of a plant, meat, seafood, poultry, eggs, dairy products, or fermentation products thereof, whose primary function in food is flavoring rather than nutritional.

The European Union's guidelines for natural flavorants are slightly different. Certain artificial flavorants are given an E number, which may be included on food labels.

Smell

Smell flavorants, or simply, flavorants, are engineered and composed in similar ways as with industrial fragrances and fine perfumes. To produce natural flavors, the flavorant must first be extractedfrom the source substance. The methods of extraction can involve solvent extraction, distillation, or using force to squeeze it out. The extracts are then usually further purified and subsequently added to food products to flavor them. To begin producing artificial flavors, flavor manufacturers must either find out the individual naturally occurring aroma compounds and mix them appropriately to produce a desired flavor or create a novel non-toxic artificial compound that gives a specific flavor.

Most artificial flavors are specific and often complex mixtures of singular naturally occurring flavor compounds combined together to either imitate or enhance a natural flavor. These mixtures are formulated by flavorists to give a food product a unique flavor and to maintain flavor consistency between different product batches or after recipe changes. The list of known flavoring agents includes thousands of molecular compounds, and the flavor chemist (flavorist) can often mix these together to produce many of the common flavors. Many flavorants consist of esters, which are often described as being "sweet" or "fruity".

Chemical	***Odor***
Diacetyl	Buttery
Isoamyl acetate	Banana
Benzaldehyde	Bitter almond
Cinnamic aldehyde	Cinnamon
Ethyl propionate	Fruity
Methyl anthranilate	Grape
Limonene	Orange
Ethyl decadienoate	Pear
Allyl hexanoate	Pineapple
Ethyl maltol	Sugar, Cotton candy
Ethylvanillin	Vanilla
Methyl salicylate	Wintergreen

The compounds used to produce artificial flavors are almost identical to those that occur naturally. It has been suggested that artificial flavors may be safer to consume than natural flavors due to the standards of purity and mixture consistency that are enforced either by the company or by law. Natural flavors in contrast may contain impurities from their sources while artificial flavors are typically more pure and are required to undergo more testing before being sold for consumption.

Flavors from food products are usually the result of a combination of natural flavors, which set up the basic smell profile of a food product while artificial flavors modify the smell to accent it.

Unlike smelling, which occurs upon inhalation, the sensing of flavors in the mouth occurs in the exhalation phase of breathing and is perceived differently by an individual. In other words, the smell of food is different depending on when you are smelling it in front of you or whether it has already entered your mouth.

Taste

While salt and sugar can technically be considered flavorants that enhance salty and sweet tastes, usually only compounds that enhance umami, as well as other secondary flavors are considered and referred to as *taste flavorants*. Artificial sweeteners are also technically flavorants.

Umami or "savory" flavorants, more commonly called taste or flavor enhancers are largely based on amino acids and nucleotides. These are typically used as sodium or calciumsalts. Umami flavorants recognized and approved by the European Union include: Certain organic and inorganic acids can be used to enhance sour tastes, but like salt and sugar these are usually not considered and regulated as flavorants under law. Each acid imparts a slightly different sour or tart taste that alters the flavor of a food.

Acid	*Description*
Glutamic acidsalts	This amino acid's sodium salt, monosodium glutamate (MSG), a notable example, is one of the most commonly used flavor enhancers in food processing. Mono and diglutamate salts are also commonly used.
Glycine salts	Simple amino acid salts typically combined with glutamic acid as flavor enhancers.
Guanylic acidsalts	Nucleotide salts typically combined with glutamic acid as flavor enhancers.
Inosinic acid salts	Nucleotide salts created from the breakdown of AMP. Due to high costs of production, typically combined with glutamic acid as flavor enhancers.
5'-ribonucleotidesalts	Nucleotide salts typically combined with other amino acids and nucleotide salts as flavor enhancers.

Acid	*Description*
Acetic acid	Gives vinegar its sour taste and distinctive smell
Ascorbic acid	Found in oranges and green peppers and gives a crisp, slightly sour taste. Better known as vitamin C.
Citric acid	Found in citrus fruits and gives them their sour taste
Fumaric acid	Not found in fruits, used as a substitute for citric and tartaric acid
Lactic acid	Found in various milk or fermented products and give them a rich tartness
Malic acid	Found in apples and gives them their sour/tart taste
Phosphoric acid	Used in all Cola drinks to give an acid taste
Tartaric acid	Found in grapes and wines and gives them a tart taste

Color

The color of food can affect flavor. For example, adding more red color to a drink increases its sweetness with darker colored solutions being rated 2–10% higher than lighter ones even though it had 1% less sucrose concentration. The effect of color is believed to be due to cognitive expectations.

Dietary Restrictions

Food manufacturers are sometimes reluctant about informing consumers about the source from where the flavor is obtained and whether it has been produced with the incorporation of substances such as animal by-products glycerin (note that glycerin is also available from vegetable sources), gelatin, and the like, and the use of alcohol in the flavors. Many Jews, Jains, Hindus, and Muslims adhere to religious dietary laws, and vegans to personal convictions, which restrict the use of animal by-products and/or alcohol in foods unless subject to oversight and inspection by their respective religious authority or less-strict or circumstantial moral belief.

In many Western countries, some consumers rely on a Jewish Kosher Pareve certification mark to indicate that natural flavorings used in a food product are free of meat and dairy (although they can still contain fish). The Vegan Society's Sunflower symbol (which is currently used by over 260 companies world wide) can also be used to see which products do not use any animal ingredients (including flavorings and colorings)

Similarly, persons with known sensitivities or allergies to food products are advised to avoid foods that contain generic "natural flavors" or to first determine the source of the flavoring before consuming the food. Such flavors may be derived from a variety of source products that are themselves common allergens, such as dairy, soy, sesame, eggs, and nuts.

Flavour Creation

Most food and beverage companies do not create their own flavors but instead employ the services of a flavor company. Food and beverage companies may require flavors for new products, product line extensions (e.g., low fat versions of existing products) or changes in formula or processing for existing products. In 2011, about U$S10.6 billion were generated with the sale of flavors; the majority of the flavors used is consumed in processed and packaged food.

The flavor creation is done by a specially trained scientist called a "flavorist". The flavorist's job combines extensive scientific knowledge of the chemical palette with artistic creativity to develop new and distinctive flavors. The flavor creation begins when the flavorist receives a brief from the client. In the brief the client will attempt to communicate exactly what type of flavor they seek, in what application it will be used, and any special requirements (e.g., must be all natural). The communication barrier can be quite difficult to overcome since most people aren't experienced at describing flavors. The flavorist will use his or her knowledge of the available chemical ingredients to create a formula and compound it on an electronic balance. The flavor will then be submitted to the client for testing. Several iterations, with feedback from the client, may be needed before the right flavor is found.

Additional work may also be done by the flavor company. For example, the flavor company may conduct sensory taste tests to test consumer acceptance of a flavor before it is sent to the client or to further investigate the "sensory space." The flavor company may also employ application specialists who work to ensure the flavor will work in the application for which it is intended. This may require special flavor delivery technologies that are used to protect the flavor during processing or cooking so that the flavor is only released when eaten by the end consumer.

Determination

Few standards are available or being prepared for sensory analysis of flavors. In chemical analysis of flavors, solid phase extraction (SPE), solid phase microextraction (SPME), and headspacegas chromatography are applied to extract and separate the flavor compounds in the sample. The determination is typically done by various mass spectrometric techniques.

Smell

The identification of primary olfactory submodalities has been attempted many times. The first serious effort at identification was made by Linnaeus, the Swedish botanist, in 1756. Over the years a large number of other classifications of odors have been made, but there is not a general agreement about primary odors in the sense of the primary tastes: sweet, sour, salty, and bitter. The human can distinguish the odors of a vast number of different molecules and describes them as aromatic, fragrant, repulsive, ethereal, resinous, spicy, burned, putrid, and so forth. Whether any of these can be considered primary is a point of debate.

Odorous substances have in common that they are either gases or volatile liquids. This is the form in which the odorant reaches the sensory epithelium, either through the nostrils with inspired air or by the back door through the mouth and throat. The receptor structures for olfaction are covered with mucus so that aqueous solubility is an asset to an odorant.

Of all the chemical elements, only 16 seem to play any role in the production of odor sensation. These are hydrogen, carbon, silicon, nitrogen, phosphorus, arsenic, antimony, bismuth, oxygen, sulfur, selenium, tellurium, fluorine, chlorine, bromine, and iodine. Only the halogens and ozone, O_3, are odorous as elements. There have been many attempts to group odorants in terms of chemical structure, with some recent success. The gross profile of the molecule, i.e., its size and shape, seem to be important. Many people have also emphasized the importance of other physicochemical features, especially the characteristics of functional groups. The olfactory system is extremely sensitive. For example, ethyl mercaptan can be detected at a concentration of 4×10^{-8} mg per liter of air. However, even at this low concentration, there would be millions of molecules in a single sniff. On the other hand, for certain odorants, a single molecule may be sufficient to excite a single olfactory receptor cell, at least in some animals. Sensitivity is no more constant in the olfactory system than in any other sensory system. It diminishes upon continuous exposure to an odor. This kind of adaptation is familiar to everyone. The smell of fried fish is not even noticed by the cook unless he leaves the kitchen and returns after some time. Odors also interact in rather specific ways. For example, adaptation to the odor of camphor results in a simultaneous increase in the threshold concentration of eucalyptus (eucalyptol) and cloves (eugenol) but not to bitter almonds (benzaldehyde). When two odorants are presented together, they can blend (violets and H_2S together yield a blend smell); they can be sensed successively; one odor can mask the other (especially if one is much stronger); or the combination can give no odor at all. The perfume industry is based on these principles.

Smell Anatomy

The olfactory receptors are located in the posterior portion of the nasal cavity in a mucosal area that measures about 5 cm^2. The location is shown in Figure 10-8, which is a sagittal section of the human nasal passages. The sensory epithelium sits at the top of the nasal cavity above the superior turbinate bones. It is yellow in color and has no synchronously beating cilia and therefore is distinguishable from the surrounding respiratory epithelia. Within this sensory region are found Bowman's glands that contain most of the yellow pigment and secrete much of the mucus that covers the sensory epithelium in a 10- to 40-m layer. The odorant must dissolve in this mucus layer in order to reach the olfactory receptors. The receptor cells are really bipolar nerve cells and therefore unlike the receptor cells of the eye, ear, and tongue which are specialized epithelial cells. The olfactory receptor cells have their nuclei in the lower 2/3 of the epithelium, but their apical ends project to the surface. The apical ends are slightly enlarged, and they possess numerous cilia that project 50 to 150 m out into the mucus. The location of the receptor sites for odorant molecules on the olfactory receptor cell is unknown. Many people believe they are on the cilia, but the existence of olfactory cells without cilia and the persistence of olfactory sensation after the cilia have been removed argue against this idea. The receptor cells are arranged in interconnected rings, one of which is illustrated in the figure. Gap junctions are thought to connect adjacent receptor cells both between spines and less frequently at their apical expansions. The centers of the rings are apparently filled with supporting cells.

At its basal end, each receptor cell gives rise to an axon that forms a part of the olfactory nerve. Olfactory nerve fibers are unmyelinated and average about 0.2 m in diameter. In the rabbit, the olfactory epithelium contains about 100 million receptor cells, and thus there are about the same number of fibers in the olfactory nerve. In man, the olfactory nerve is short, merely perforating the cribriform plate into the brain cavity and terminating in the olfactory bulb. Olfactory nerve fibers converge onto mitral cells in the bulb, which send their axons (about 100,000 of them) in the olfactory tract to the piriform cortex, the periamygdaloid area, and the olfactory tubercle. The olfactory bulb contains interneurons, the granule cells that mediate a form of lateral inhibition between mitral cells; an active mitral cell inhibits (reduces) activity in its neighbors. There are also centrifugal fibers to the olfactory bulb that primarily mediate inhibition of mitral cell activity. The functions of this connection and of the lateral inhibition are not known.

Unlike other sensory pathways to the cerebral cortex, the olfactory pathway does not relay in the thalamus as we have just seen. However, fibers do leave the olfactory cortical areas and relay in the thalamus on their way to the hypothalamus or other areas where they perhaps play a role in the regulation of the intake of food and other behaviors that depend upon olfactory information.

Smell Physiology

The coding mechanism for olfactory quality (submodality) has been just as elusive as the identification of the qualities themselves. Microelectrode recordings from the olfactory epithelium yield two sorts of responses to olfactory stimulation. Both can be seen in Figure 10-10. First, there is a slow-wave potential reminiscent of the generator potentials of mechanoreceptors (see Chapter 4). However, this potential is much larger than generator potentials recorded extracellularly from single receptors elsewhere. It also does not have a fixed relationship to the time of spike initiation, as can be seen in the figure. This potential probably is the summation of the generator potentials of a number of nearby receptor cells and perhaps potentials generated by supporting cells.

Second, there are the familiar spike discharges in the receptor cell axons or somata. It can be seen from this typical example that receptors are not highly specific in their responses to odorants. This particular cell responded to camphor, limonene, carbon disulfide, ethyl butyrate, and musk xylene, but it was insensitive to nitrobenzene, benzaldehyde, n-butanol and pyridine. In general, each olfactory receptor responds to a variety of odorants, suggesting a very complex ensemble code.

The same sort of behavior has also been seen in mitral cells of the olfactory bulb. Perhaps, like the gustatory system, the olfactory system signals smell quality by the pattern of activity in the ensemble of olfactory receptors. The larger number of odorants compared to taste stimuli makes this hypothesis difficult to test. Studies have shown that odors may be encoded in terms of the position in the epithelium of responding receptors and the relative timing of discharges across the sheet of receptors. Spatial-temporal patterning is the basis of chromatographic separation of molecules; a similar process might be used by the olfactory epithelium in odor discrimination. Compounds with similar smells have been shown to be similar in terms of these two parameters. As yet, the evidence favoring this hypothesis must be viewed as preliminary.

The sensations of taste are mediated by taste buds located in papillae of the tongue. There are four taste submodalities: sweet, salty, sour and bitter. The four submodalities are not completely independent, but interact in successive tastings. Many taste stimuli (except quinine, sucrose, and NaCl) give different sensations at different concentrations. Taste receptors normally respond to more than one of the taste submodalities; thus, taste quality is probably encoded as an ensemble code. Taste intensity is probably encoded in terms of the total number of impulses (or frequency) discharged in all the fibers sensitive to a particular substance. Substances that taste the same tend to excite taste receptors in a similar way. Olfactory sensations are evoked by volatile substances, but primary submodalities have not been identified. Olfactory receptors respond to a variety of odorants, and thus some kind of ensemble code is likely for this modality as well.

Our Nose Knows Ten Types of Odour

Most of us know that there are five taste receptors. Now a US study has determined that smells can be lumped into one of ten categories which include citrus, mint, pungent and decaying.

The findings are published today in the open access journal *PLOS ONE*.

Researchers took a sample of 144 smell profiles from theAtlas of Odor Character Profiles - a 1985 publication considered as the 'gold standard' - and applied statistical analysis to look for commonalties among the various odours.

"It's an open question how many fundamental types of odour qualities there are," says study co-author Jason Castro of Bates College, Maine. "This is in striking contrast to olfaction's 'sister sense', taste, where we know that five basic qualities seem to organize sensations - [sweet, sour, salt, bitter and umami]."

The team identified 10 basic odour clusters. While some can easily be labelled, such as fragrant, woody or chemical, others are more complicated. For example, one cluster includes minty, aniseed, spice, eucalyptus and camphor; while another lists popcorn, smoky, nutty, almond, oily and mouldy.

Professor David O'Carroll, a sensory neuroscientist at theUniversity of Adelaide who was not involved in the study, says it's a well-constructed piece of research.

"What they have done for the first time is use a clever mathematical technique to cluster these different odours into ten groups that are related to one another in terms of the chemical that stimulate them," O'Carroll told ABC Breakfast.

He says understanding how we detect odour is much more complex than sight or smell.

"Understanding how we perceive colour is simple because we've only got three different types of colour receptors in our eye," says O'Carroll.

"With odour, it's potentially frighteningly complicated because you're dealing with a ten-dimensional odour space. Making sense of that has been something that is very challenging."

Predicting Smells

The US researchers are now aiming to understand smell from the opposite direction, looking at chemical structures in an attempt to predict how it will smell.

"That's something that nobody's really done with any kind of compelling accuracy," says Castro. "And obviously perfume companies, flavour and fragrance companies, are really interested in doing that well."

O'Connell agrees: "If you can understand and predict what something is going to smell like to a human, then perhaps you can develop that perfume that is going to take the world by storm."

Chapter 11

Determination of Carbohydrates

Carbohydrates are one of the most important components in many foods. Carbohydrates may be present as isolated molecules or they may be physically associated or chemically bound to other molecules. Individual molecules can be classified according to the number of monomers that they contain as monosaccharides, oligosaccharides or polysaccharides. Molecules in which the carbohydrates are covalently attached to proteins are known as *glycoproteins*, whereas those in which the carbohydrates are covalently attached to lipids are known as *glycolipids*. Some carbohydrates are digestible by humans and therefore provide an important source of energy, whereas others are indigestible and therefore do not provide energy. Indigestible carbohydrates form part of a group of substances known as *dietary fiber*, which also includes lignin. Consumption of significant quantities of dietary fiber has been shown to be beneficial to human nutrition, helping reduce the risk of certain types of cancer, coronary heart disease, diabetes and constipation. As well as being an important source of energy and dietary fiber, carbohydrates also contribute to the sweetness, appearence and textural characteristics of many foods. It is important to determine the type and concentration of carbohydrates in foods for a number of reasons.

- *Standards of Identity* - foods must have compositions which conform to government regulations
- *Nutritional Labeling* - to inform consumers of the nutritional content of foods
- *Detection of Adulteration* - each food type has a carbohydrate "fingerprint"
- *Food Quality* - physicochemical properties of foods such as sweetness, appearance, stability and texture depend on the type and concentration of carbohydrates present.
- *Economic* - industry doesn't want to give away expensive ingredients

- *Food Processing* - the efficiency of many food processing operations depends on the type and concentration of carbohydrates that are present

Classification of Carbohydrates

Monosaccharides

Monosaccharides are water-soluble crystalline compounds. They are aliphatic aldehydes or ketones which contain one carbonyl group and one or more hydroxyl groups. Most natural monosachharides have either five (pentoses) or six (hexoses) carbon atoms. Commonly occurring hexoses in foods are glucose, fructose and galactose, whilst commonly occurring pentoses are arabinose and xylose. The reactive centers of monosaccharidesare the carbonyl and hydroxyl groups.

Oligosaccharides

These are relatively low molecular weight polymers of monosaccharides (< 20) that are covalently bonded through glycosidic linkages. Disaccharides consist of two monomers, whereas trisaccharides consist of three. Oligosaccharides containing glucose, fructose and galactose monomers are the most commonly occurring in foods.

Polysaccharides

The majority of carbohydrates found in nature are present as polysaccharides. Polysaccharides are high molecular weight polymers of monosaccharides (> 20). Polysaccharides containing all the same monosaccharides are called homopolysaccharides (*e.g.*, starch, cellulose and glycogen are formed from only glucose), whereas those which contain more than one type of monomer are known as heteropolysaccharides (*e.g.*, pectin,hemicellulose and gums).

Methods of Analysis

A large number of analytical techniques have been developed to measure the total concentration and type of carbohydrates present in foods (see *Food Analysis* by Nielssen or *Food Analysis* by Pomeranz and Meloan for more details). The carbohydrate content of a food can be determined by calculating the percent remaining after all the other components have been measured: %carbohydrates = 100 - %moisture - %protein - %lipid - %mineral. Nevertheless, this method can lead to erroneous results due to experimental errors in any of the other methods, and so it is usually better to directly measure the carbohydrate content for accurate measurements

Monosaccharides and Oligosaccharides

Sample Preparation

The amount of preparation needed to prepare a sample for carbohydrate analysis depends on the nature of the food being analyzed. Aqueous solutions, such as fruit juices, syrups and honey, usually require very little preparation prior to analysis. On the other hand, many foods contain carbohydrates that are physically associated or chemically bound to other components, *e.g.*, nuts, cereals, fruit, breads and vegetables. In these foods it is usually necessary to isolate the carbohydrate from the rest of the food before it can be analyzed. The precise method of carbohydrate isolation depends on the carbohydrate type, the food matrix type and the purpose ofanalysis, however, there are some procedures that are common to many isolation techniques. For example, foods are usually dried under vacuum (to prevent thermal degradation), ground to a fine powder (to enhance solvent extraction) and then defatted by solvent extraction.

One of the most commonly used methods of extracting low molecular weight carbohydrates from foods is to boil a defatted sample with an 80% alcohol solution. Monosaccharides and oligosaccharides are soluble in alcoholic solutions, whereas proteins, polysaccharides and dietary fiber are insoluble. The soluble components can be separated from the insoluble components by filtering the boiled solution and collecting the filtrate (the part which passes through the filter) and the retentante (the part retained by the filter). These two fractions can then be dried and weighed to determine their concentrations. In addition, to monosaccharides and oligosaccharides various other small molecules may also be present in the alcoholic extract that could interfere with the subsequent analysis *e.g.,* amino acids, organic acids, pigments, vitamins, minerals *etc.* It is usually necessary to remove these components prior to carrying out a carbohydrate analysis. This is commonly achieved by treating the solution with clarifying agents or by passing it through one or more ion-exchange resins.

- *Clarifying agents.* Water extracts of many foods contain substances that are colored or produce turbidity, and thus interfere with spectroscopic analysis or endpoint determinations. For this reason solutions are usually *clarified* prior to analysis. The most commonly used clarifying agents are heavy metal salts (such as lead acetate) which form insoluble complexes with interfering substances that can be removed by filtration or centrifugation. However, it is important that the clarifying agent does not precipitate any of the carbohydrates from solution as this would cause an underestimation of the carbohydrate content.
- *Ion-exchange.* Many monosaccharides and oligosaccharides are polar non-charged molecules and can therefore be separated from charged molecules by passing samples through ion-exchange columns. By using a combination of a positively and a negatively charged column it is possible to remove most charged contaminants. Non-polar molecules can be removed by passing a solution through a column with a non-polar stationary phase. Thus proteins, amino acids, organic acids, minerals and hydrophobic compounds can be separated from the carbohydrates prior to analysis.

Prior to analysis, the alcohol can be removed from the solutions by evaporation under vacuum so that an aqueous solution of sugars remains.

Chromatographic and Electrophoretic Methods

Chromatographic methods are the most powerful analytical techniques for the analysis of the *type* and *concentration* of monosaccharides and oligosaccharides in foods. Thin layer chromatography (TLC), Gas chromatography (GC) and High Performance Liquid chromatography (HPLC) are commonly used to separate and identify carbohydrates. Carbohydrates are separated on the basis of their differential adsorption characteristics by passing the solution to be analyzed through a column. Carbohydrates can be separated on the basis of their partition coefficients, polarities or sizes, depending on the type of column used. HPLC is currently the most important chromatographic method for analyzing carbohydrates because it is capable of rapid, specific, sensitive and precise measurements. In addition, GC requires that the samples be volatile, which usually requires that they be derivitized, whereas in HPLC samples can often be analyzed directly. HPLC and GC are commonly used in conjunction with NMR or mass spectrometry so that the chemical structure of the molecules that make up the peaks can also be identified.

Carbohydrates can also be separated by electrophoresis after they have been derivitized to make them electrically charged, *e.g.*, by reaction with borates. A solution of the derivitized carbohydrates is applied to a gel and then a voltage is applied across it. The carbohydrates are then separated on the basis of their size: the smaller the size of a carbohydrate molecule, the faster it moves in an electrical field.

Chemical Methods

A number of chemical methods used to determine monosaccharides and oligosaccharides are based on the fact that many of these substances are *reducing agents* that can react with other components to yield precipitates or colored complexes which can be quantified. The concentration of carbohydrate can be determined gravimetrically, spectrophotometrically or by titration. Non-reducing carbohydrates can be determined using the same methods if they are first hydrolyzed to make them reducing. It is possible to determine the concentration of both non-reducing and reducing sugars by carrying out an analysis for reducing sugars before and afterhydrolyzation. Many different chemical methods are available for quantifying carbohydrates. Most of these can be divided into three catagories: titration, gravimetric and colorimetric. An example of each of these different types is given below.

Titration Methods

The Lane-Eynon method is an example of a tritration method of determining the concentration of reducing sugars in a sample. A burette is used to add the carbohydrate solution being analyzed to a flask containing a known amount of boiling copper sulfate solution and a methylene blue indicator. The reducing sugars in the carbohydrate solution react with the copper sulfate present in the flask. Once all the copper sulfate in solution has reacted, any further addition of reducing sugars causes the indicator to change from blue to white. The volume of sugar solution required to reach the end point is recorded. The reaction is not stoichemetric, which means that it is necessary to prepare a calibration curve by carrying out the experiment with a series of standard solutions of known carbohydrate concentration.

The disadvantages of this method are (i) the results depend on the precise reaction times, temperatures and reagent concentrations used and so these parameters must be carefully controlled; (ii) it cannot distinguish between different types of reducing sugar, and (iii) it cannot directly determine the concentration of non-reducing sugars, (iv) it is sucseptible to interference from other types of molecules that act as reducing agents..

Gravimetric Methods

The Munson and Walker method is an example of a gravimetric method of determining the concentration of reducing sugars in a sample. Carbohydrates are oxidized in the presence of heat and an excess of copper sulfate and alkaline tartrate under carefully controlled conditions which leads to the formation of a copper oxide precipitate:

reducing sugar + Cu^{2+} + base □ oxidized sugar + CuO_2 (precipitate)

The amount of precipitate formed is directly related to the concentration of reducing sugars in the initial sample. The concentration of precipitate present can be determined gravimetrically (by filtration, drying and weighing), or titrimetrically (by redissolving the precipitate and titrating with a suitable indicator). This method suffers from the same disadvantages as the Lane-Eynon method, neverthless, it is more reproducible and accurate.

Colorimetric Methods

The Anthrone method is an example of a colorimetric method of determining the concentration of the total sugars in a sample. Sugars react with the anthrone reagent under acidic conditions to yield a blue-green color. The sample is mixed with sulfuric acid and the anthrone reagent and then boiled until the reaction is completed. The solution is then allowed to cool and its absorbance is measured at 620 nm. There is a linear relationship between the absorbance and the amount of sugar that was present in the original sample. This method determines both reducing and non-reducing sugars because of the presence of the strongly oxidizing sulfuric acid. Like the other methods it is non-stoichemetric and therefore it is necessary to prepare a calibration curve using a series of standards of known carbohydrate concentration.

The Phenol - Sulfuric Acid method is an example of a colorimetric method that is widely used to determine the total concentration of carbohydrates present in foods. A clear aqueous solution of the carbohydrates to be analyzed is placed in a test-tube, then phenol and sulfuric acid are added. The solution turns a yellow-orange color as a result of the interaction between the carbohydrates and the phenol. The absorbance at 420 nm is proportional to the carbohydrate concentration initially in the sample. The sulfuric acid causes all non-reducing sugars to be converted to reducing sugars, so that this method determines the total sugars present. This method is non-stoichemetric and so it is necessary to prepare a calibration curve using a series of standards of known carbohydrate concentration.

Enzymatic Methods

Analytical methods based on enzymes rely on their ability to catalyze specific reactions. These methods are rapid, highly specific and sensitive to low concentrations and are therefore ideal for determination of carbohydrates in foods. In addition, little sample preparation is usually required. Liquid foods can be tested directly, whereas solid foods have to be dissolved in water first. There are many enzyme assay kits which can be purchased commercially to carry out analysis for specific carbohydrates. Manufacturers of these kits provide detailed instructions on how to carry out the analysis. The two methods most commonly used to determine carbohydrate concentration are: (i) allowing the reaction to go to completion and measuring the concentration of the product, which is proportional to the concentration of the initial substrate; (ii). measuring the initial rate of the enzyme catalyzed reaction because the rate is proportional to the substrate concentration. Some examples of the use of enzyme methods to determine sugar concentrations in foods are given below:

D-Glucose/D-Fructose

This method uses a series of steps to determine the concentration of both glucose and fructose in a sample. First, glucose is converted to glucose-6-phosphate (G6P) by the enzyme hexakinase and ATP. Then, G6P is oxidized by $NADP^+$ in the presence of G6P-dehydrogenase (G6P-DH)

G6P + $NADP^+$ □ gluconate-6-phosphate + NADPH + H^+

The amount of NADPH formed is proportional to the concentration of G6P in the sample and can be measured spectrophotometrically at 340nm. The fructose concentration is then determined by converting the fructose into glucose, using another specific enzyme, and repeating the above procedure.

Maltose/Sucrose

The concentration of maltose and sucrose (disaccharides) in a sample can be determined after the concentration of glucose and fructose have been determined by the previous method. The maltose and sucrose are broken down into their constituent monosaccharides by the enzyme a-glucosidase:

maltose + H_2O 2 glucose

sucrose +H_2O □ glucose + fructose

The concentrations of glucose and fructose can then be determined by the previous method. The major problem with this method is that many other oligosaccharides are also converted to monosaccharides by a-glucosidase, and it is difficult to determine precisely which oligosaccharides are present. This method is therefore useful only when one knows the type of carbohydrates present, but not their relative concentrations. Various other enzymatic methods are available for determining the concentration of other monosaccharides and oligosaccharides, *e.g.*, lactose, galactose and raffinose (see *Food Analysis* Nielssen).

Physical Methods

Many different physical methods have been used to determine the carbohydrate concentration of foods. These methods rely on their being a change in some physicochemical characteristic of a food as its carbohydrate concentration varies. Commonly used methods include polarimetry, refractive index, IR, and density.

Polarimetry

Molecules that contain an asymmetric carbon atom have the ability to rotate plane polarized light. A polarimeter is a device that measures the angle that plane polarized light is rotated on passing through a solution. Apolarimeter consists of a source of monochromatic light, a polarizer, a sample cell of known length, and an analyzer to measure the angle of rotation. The extent of polarization is related to the concentration of the optically active molecules in solution by the equation , where a is the measured angle of rotation, [a] is the optical activity (which is a constant for each type of molecule), l is the pathlength and c is the concentration. The overall angle of rotation depends on the temperature and wavelength of light used and so these parameters are usually standardized to 20°C and 589.3 nm (the D-line for sodium). A calibration curve of a versus concentration is prepared using a series of solutions with known concentration, or the value of [a] is taken from the literature if the type of carbohydrates present is known. The concentration of carbohydrate in an unknown sample is then determined by measuring its angle of rotation and comparing it with the calibration curve.

Refractive Index

The refractive index (n) of a material is the velocity of light in a vacuum divided by the velocity of light in the material ($n = c/c_m$). The refractive index of a material can be determined by measuring the angle of refraction (r) and angle of incidence (i) at a boundary between it and another material of known refractive index (Snell's Law: $\sin(i)/\sin(r) = n_2/n_1$). In practice, the refractive index of carbohydrate solutions is usually measured at a boundary with quartz.The refractive index of a carbohydrate solution increases with increasing concentration and so can be used to measure the amount of carbohydrate present. The RI is also temperature and wavelength dependent and so measurements are usually made at a specific temperature (20 °C) and wavelength (589.3nm). This method is quick and simple to carry out and can be performed with simple hand-held instruments. It is used routinely in industry to determine sugar concentrations of syrups, honey, molasses, tomato products and jams.

Density

The density of a material is its mass divided by its volume. The density of aqueous solutions increases as the carbohydrate concentration increases. Thus the carbohydrate concentration can be determined by measuring density, *e.g.*, using density bottles or hydrometers. This technique is routinely used in industry for determination of carbohydrate concentrations of juices and beverages.

Infrared

A material absorbs infrared due to vibration or rotation of molecular groups. Carbohydrates contain molecular groups that absorb infrared radiation at wavelengths where none of the other major food constituents absorb consequently their concentration can be determined by measuring the infrared absorbance at these wavelengths. By carrying out measurements at a number of different specific wavelengths it is possible to simultaneously determine the concentration of carbohydrates, proteins, moisture and lipids. Measurements are normally carried out by measuring the intensity of an infrared wave reflected from the surface of a sample: the greater the absorbance, the lower the reflectance. Analytical instruments based on infrared absorbance are non-destructive and capable of rapid measurements and are therefore particularly suitable for on-line analysis or for use in a quality control laboratory where many samples are analyzed routinely.

More sophisticated instrumental methods are capable of providing information about the molecular structure of carbohydrates as well as their concentration, *e.g.*, NMR or mass spectrometry.

Immunoassays

Immuoassays are finding increasing use in the food industry for the qualitative and quantitative analysis of food products. Immunoassays specific for low molecular weight carbohydrates are developed by attaching the carbohydrate of interest to a protein, and then injecting it into an animal. With time the animal develops antibodies specific for the carbohydrate molecule. These antibodies can then be extracted from the animal and used as part of a test kit for determining the concentration of the specific carbohydrate in foods. Immuoassays are extremely sensitive, specific, easy to use and rapid.

Analysis of Polysaccharides and Fiber

A wide variety of polysaccharides occur in foods. Polysaccharides can be classified according to their molecular characteristics (*e.g.*, type, number, bonding and sequence of monosaccharides), physicochemical characteristics (*e.g.*, water solubility, viscosity, surface activity) and nutritional function (*e.g.*, digestible or non-digestible). Most polysaccharides contain somewhere between 100 and several thousand monosaccharides. Some polysaccharides contain all the same kind of monosaccharide (homopolysaccharides), whereas others contain a mixture of different kinds of monosaccharide (heteropolysaccharides). Some polysaccharides exist as linear chains, whereas others exist as branched chains. Some polysaccharides can be digested by human beings and therefore form an important source of energy (*e.g.*, starch), whereas others are indigestible (*e.g.*, cellulose,hemicellulose and pectins). These indigestible polysaccharides form part of a group of substances known as *dietary fiber*, which also includes lignin (which is a polymer of aromatic molecules). Consumption of many types of dietary fiber has been shown to have beneficial physiologically functional properties for humans, *e.g.*, prevention of cancer, heart disease and diabetes.

Analysis of Starch

Starch is the most common digestible polysaccharide found in foods, and is therefore a major source of energy in our diets. In its natural form starch exists as water-insoluble granules (3 - 60 mm), but in many processed foods the starch is no longer in this form because of the processing treatments involved (*e.g.*, heating).◎ It consists of a mixture of two glucose homopolysaccharides: *amylose* (500-2000 glucose units) which is linear, and*amylopectin* (>1,000,000 glucose units) which is extensively branched. These two kinds of starch have different physiochemical properties and so it is often important to determine the concentration of each individual component of the starch, as well as the overall starch concentration.

- *Sample preparation.* The starch content of most foods cannot be determined directly because the starch is contained within a structurally and chemically complex food matrix. In particular, starch is often present in a semi-crystalline form (granular or retrograded starch) that is inaccessible to the chemical reagents used to determine its concentration. It is therefore necessary to isolate starch from the other components present in the food matrix prior to carrying out a starch analysis.

In natural foods, such as legumes, cereals or tubers, the starch granules are usually separated from the other major components by drying, grinding, steeping in water, filtration and centrifugation. The starch granules are water-insoluble and have a relatively high density (1500 kg/m^3) so that they will tend to move to the bottom of a container during centrifugation, where they can be separated from the other water-soluble and less dense materials. Processed food samples are normally dried, ground and then dispersed in hot 80% ethanol solutions. The monosaccharides and oligosaccharides are soluble in the ethanol solution, while the starch is insoluble. Hence, the starch can be separated from the sugars by filtering or centrifuging the solution. If any semi-crystalline starch is present, the sample can be dispersed in water and heated to a temperature where the starch gelatinizes (> 65 °C). Addition of perchloric acid or calcium chloride to the water prior to heating facilitates the solubilization of starches that are difficult to extract.

- *Analysis methods.* Once the starch has been extracted there are a number of ways to determine its concentration:

Specific enzymes are added to the starch solution to breakdown the starch to glucose. The glucose concentration is then analyzed using methods described previously (*e.g.*, chromatography or enzymatic methods). The starch concentration is calculated from the glucose concentration.

Iodine can be added to the starch solution to form an insoluble starch-iodine complex that can be determined *gravimetrically* by collecting, drying and weighing the precipitate formed or *titrimetrically* by determining the amount of iodine required to precipitate the starch.

If there are no other components present in the solution that would interfere with the analysis, then the starch concentration could be determined using physical methods, *e.g.*, density, refractive index or polarimetry.

The amylose and amylopectin concentrations in a sample can be determined using the same methods as described for starch once the amylose has been separated from the amylopectin. This can be achieved by adding chemicals that form an insoluble complex with one of the components, but not with the other, *e.g.* some alcohols precipitate amylose but not amylopectin. Some of the methods mentioned will not determine the concentration of resistant starch present in the sample. If the concentration of resistant starch is required then an additional step can be added to the procedure where dimethylsulfoxide (DMSO) is added to dissolve the resistant starch prior to carrying out the analysis.

Analysis of Fibers

Over the past twenty years or so nutritionists have become aware of the importance of fiber in the diet. Liberal consumption of fiber helps protect against colon cancer, cardiovascular disease and constipation. Adequate intake of dietary fiber is therefore beneficial to good health. Dietary fiber is defined as plant polysaccharides that are indigestible by humans, plus lignin. The major components of dietary fiber are cellulose, hemicellulose, pectin, hydrocolloids and lignin. Some types of starch, known as *resistant starch,* are also indigestible by human beings and may be analyzed as dietary fiber. The basis of many fiber analysis techniques is therefore to develop a procedure that mimics the processes that occur in the human digestive system.

Major Components of Dietary Fiber

Cell Wall Polysaccharides

Cellulose occurs in all plants as the principal structural component of the cell walls, and is usually associated with various hemicelluloses and lignin. The type and extent of these associations determines the characteristic textural properties of many edible plant materials. Cellulose is a long linear homopolysaccahride of glucose, typically having up to 10,000 glucose subunits. Cellulose molecules aggregate to form microfibrils that provide strength and rigidity in plant cell walls. Hemicelluloses are a heterogeneous group of branched heteropolysaccharides that contain a number of different sugars in their backbone and side-chains. By definition hemicelluloses are soluble in dilute alkali solutions, but insoluble in water. Pectins are another form of heteropolysaccharides found in cell walls that are rich in uronic acids, soluble in hot water and that are capable of forming gels.

Non Cell Wall Polysaccharides

This group of substances are also indigestible carbohydrates, but they are not derived from the cell walls of plants. Non cell wall polysaccharides include hydrocolloids such as guar and locust bean gum, gum arabic, agar, alginates and caragenans which are commonly used in foods as gelling agents, stabilizers and thickeners.

Lignin

Lignin is a non-carbohydrate polymer that consists of about 40 aromatic subunits which are covalently linked. It is usually associated with cellulose and hemicelluloses in plant cell-walls.

Common Procedures in Sample Preparation and Analysis

There are a number of procedures that are commonly used in many of the methods for dietary fiber analysis:

- *Lipid removal.* The food sample to be analyzed is therefore dried, ground to a fine powder and then the lipids are removed by solvent extraction.
- *Protein removal.* Proteins are usually broken down and solubilized using enzymes, strong acid or strong alkali solutions. The resulting amino acids are then separated from insoluble fiber by filtration or from total fiber by selective precipitation of the fiber with ethanol solutions.
- *Starch removal.* Semi-crystalline starch is gelatinized by heating in the presence of water, and then the starch is broken down and solubilized by specific enzymes, strong acid or strong alkali. The glucose is then separated from insoluble fiber by filtration or separated from total fiber by selective precipitation of the fiber with ethanol solutions.
- *Selective precipitation of fibers.* Dietary fibers can be separated from other components in aqueous solutions by adding different concentrations of ethanol to cause selective precipitation. The solubility ofmonosaccharides, oligosaccharides and polysaccharides depends on the ethanol concentration. *Water*: monosaccharides, oligosaccharides, some polysaccharides and amino acids are soluble; other polysaccharides and fiber are insoluble. *80% ethanol solutions:* monosaccharides, oligosaccharides and amino acids are soluble; polysaccharides and fibers are insoluble. For this reason, *concentrated ethanol solutions* are often used to selectively precipitate fibers from other components.
- *Fiber analysis.* The fiber content of a food can be determined either *gravimetrically* by weighing the mass of an insoluble fiber fraction

isolated from a sample or *chemically* by breaking down the fiber into its constituent monosaccharides and measuring their concentration using the methods described previously.

Crude Fiber Method

The crude fiber method gives an estimate of indigestible fiber in foods. It is determined by sequential extraction of a defatted sample with 1.25% H_2SO_4 and 1.25% NaOH. The insoluble residue is collected by filtration, dried, weighed and ashed to correct for mineral contamination of the fiber residue. Crude fiber measures cellulose and lignin in the sample, but does not determine hemicelluloses, pectins and hydrocolloids, because they are digested by the alkali and acid and are therefore not collected. For this reason many food scientists believe that its use should be discontinued. Nevertheless, it is a fairly simple method to carry out and is the official AOAC method for a number of different foodstuffs.

Total, Insoluble and Soluble Fiber Method

The basic principle of this method is to isolate the fraction of interest by selective precipitation and then to determine its mass by weighing. A gelatinized sample of dry, defatted food is enzymatically digested with a-amylase, amyloglucosidase and protease to break down the starch and protein components. The total fiber content of the sample is determined by adding 95% ethanol to the solution to precipitate all the fiber. The solution is then filtered and the fiber is collected, dried and weighed. Alternatively, the water-soluble and water-insoluble fiber components can be determined by filtering the enzymatically digested sample. This leaves the *soluble* fiber in the filtrate solution, and the *insoluble* fiber trapped in the filter. The insoluble component is collected from the filter, dried and weighed. The soluble component is precipitated from solution by adding 95% alcohol to the filtrate, and is then collected by filtration, dried and weighed. The protein and ash content of the various fractions are determined so as to correct for any of these substances which might remain in the fiber: Fiber = residue weight - weight of (protein + ash).

This method has been officially sanctioned by the AOAC and is widely used in the food industry to determine the fiber content of a variety of foods. Its main disadvantage is that it tends to overestimate the fiber content of foods containing high concentrations of simple sugars, *e.g.*, dried fruits, possibly because they get trapped in the precipitates formed when the ethanol is added.

Chemical Methods

In chemical methods, the fiber content is equal to the sum of all nonstarch monosaccharides plus lignin remaining once all the digestible carbohydrates have been removed. Monosaccharides are measured using the various methods described previously.

Englyst-Cummings Procedure

A defatted food sample is heated in water to gelatinize the starch. Enzymes are then added to digest the starch and proteins. Pure ethanol is added to the solution to precipitate the fiber, which is separated from the digest by centrifugation, and is then washed and dried. The fiber is then hydrolyzed using a concentrated sulfuric acid solution to break it down into its constituent monosaccharides, whose concentration is determined using the methods described previously, *e.g.*, colorimetrically or chromatographically. The mass of fiber in the original sample is assumed to be equal to the total mass of monosaccharides present. The concentration of insoluble and soluble dietary fiber can also be determined by this method, using similar separation steps as for the *total, insoluble and soluble* gravimetric method mentioned above.

This method can be used to determine the total, soluble and insoluble fiber contents of foods, but does not provide information about the *lignin* content. This is because lignin is not a polysaccharide, and so it is not broken down to monosaccharides during the acid digestion. For most foods this is not a problem because they have low lignin concentrations anyway. If a food does contain significant amounts of lignin then another method should be used, *e.g.*, the gravimetric method or more sophisticated chemical methods (*e.g.*, the Theander-Marlett method).

Bibliography

ADM. 1992. Food applications of citric acid and its salts. ADM Technical Bulletin. Decatur, III., USA

ANON. 1960. Food Composition Tables. Rome: FAO.

ANON. 1993. Cherry and apricot oils are safe for use. In Lipid Technology. May/ June 1993, pp. 53-54.

BIAUGEAUD, H. 1994. Food processing equipment. Technical Bulletin. Henri Biaugeaud, S.A.: Arcueil.

BISWAL, R.N. and BOZORGMEHR, K. 1991. Equilibrium data for osmotic concentration of potato in NaCl-water solution. In Journal of Food Process Engineering, 14, pp. 237-245.

BOLIN, H.R. and STAFFORD, A.E. 1974. Effect of processing and storage on provitamin A and vitamin C in apricots. In Journal of Food Science. Vol. 39, pp. 10341035.

BONGIRWIR, D.R. and Sreenivasan, A. 1977. Studies on osmotic dehydration of bananas. In J. Food Sci. Technol. India. 14, pp. 104-112.

C.L.V. 1994. Comptoir Lyonnais de verreries. Technical bulletin. C.L.V.: Villeurbaine.

CHAN, H.T. Jr. 1984. (Editor). Handbook of Tropical Foods. New York/Basel: Marcel Dekker.

COLIN, D. 1992. Recent trends in fruit and vegetable processing. In Food Science and Technology Today, 7, (2)), pp. 111 - 116.

DESROSIER, N.W. and DESROSIER, J.N. 1977. Technology of Food Preservation, 4th Edition, AVI Publishing Co. Westport, Conn.

ELLIS, R.F. 1963. Metal containers for food. In Food Processing Operations, Vol. 2. M.A. Joslin and J.L. Heid (Editors). AVI Publishing Co., Westport, Conn.

FAO. 1968. Fruits et legumes en Afrique occidentale. Rome:FAO.

FAO. 1969a. Sun drying fruits and vegetables in Sudan. Jackson, T.H. and Mohammed, B.B. FAO Report. Rome: FAO.

FAO. 1969b. Sun drying of fruits and vegetables. Jackson, T.H. and Mohammed, B.B. Agricultural Services Bulletin No. 5. Rome: FAO.

FAD/WFP. 1970. Food Storage Manual, Part II. Food and Commodities. Ashman, F. (Editor). Rome: FAO.

FAO. 1972. Fruits tropicaux transformes. Rome: FAO.

FAO. 1973. Fruit juice processing. Agricultural Services Bulletin No. 13. Rome: FAO.

FAO. 1986a. Solar drying in ASEAN countries. RAPA publications 1986/27. Rome: FAO.

FAO. 1986b. Prevention of post-harvest food losses. Rome: FAO.

FAO. 1988a. Root and tuber crops, plantains and bananas in developing countries. FAO Plant Production and Protection Paper No. 87. Rome: FAO.

FAO. 1988b. Manuals of Food Quality Control. FAO Food and Nutrition Paper. No. 14. Rome: FAO.

FAO. 1988c. Traditional food plants. FAO Food and Nutrition Paper No. 42. Rome: FAO.

FAO. 1990a. Rural processing and preserving techniques for fruits and vegetables. Rome: FAO.

FAO. 1990b. Comment conserver less tomates -Trots techniques de transformation et de conservation artisanale. Rome: FAO.

FAO. 1991. Post-harvest and processing technologies of African staple foods: a technical compendium. Agricultural Services Bulletin No. 89. Rome: FAO.

FAO. 1992a. Production Yearbook 1991. Rome: FAO.

FAO. 1992b. Trade Yearbook 1991. Rome: FAO.

FAO. 1992c. The marvellous mango bar. Amoriggi, G. In Ceres. July/Aug. pp. 25-28. Rome: FAO.

FARKAS, D.F. 1991. How to describe a process for quality control. PA'91 - Session B-2, Oregon State University, Corvallis, Or.

FDA. 1991. Code of Federal Regulations. 7 CFR 52. Food products.

FDA. 1986. Code of Federal Regulations. 21 CFR 110. Current Good Manufacturing Practice in Manufacturing, Packaging or Holding Human Foods.

FEINBERG, B. 1973. "Vegetables". In Food Dehydration, 2nd Edition, Vol. 2. W.B. Van Arsdel, M.J. Copley and A.I. Morgan, Jr.(Editors). AVI Publishing Co., Westport, Conn.

DA FONSECA, J.L.F. and MOY, J.H. 1975. A processing plant for tomato and tropical fruit products. University of Hawaii.

GARROTTE, R.L., SILVA, E. and BERTONE, R.A. Osmotic concentration at 5 and 25°C of pear and apple cubes and strawberry halves. Lebensm.-Wiss.u.-Technol. 25. pp. 133-138.

GRATIAS, C. 1994. Technical Bulletin - Food industry equipment. C. Gratias & Cie.: Orly.

HANKE, H. 1992. The use of fruit juices in confectionery products.In 46th P.M.C.A. Production Conference, 1992. pp.98-100.

HANSELMANN, A. 1950. Praktisches Handbuch des Marmeladenkochers. Braunschweig: Dr. Serger & Hempel.

HARRIS, R.S. and KARMAS, E. 1975. Nutritional Evaluation of Food Processing. 2nd Edition.AVI Publishing Co.,Westport,Conn.

HEISS,R. 1955.Forschritte in der Technologie des Konservierens von Gemuse und Obst.Braunschweig: Dr. Serger & Hempel.

HERRMANN, K. 1969. Gemuse und Gemusedauerwaren. Berlin & Hamburg: Paul Parey.

HOECHST. 1994. Technical Bulletin SunettR. Hoechst, A.G.

HORWICH ALLEN, A. 1992. Seeds of Success:Designing Foods using Vegetables. Food Product Design. Dec. 1992. pp.61-62, 65-66.

INGEGNO, C. 1992. A fresh look at dried fruit. In Food Product Design, Oct. 1992, pp. 48, 50.

Keyle, R.h., Gresham. W.a., Jr. and COLLUM, Ch.E. 1964. Petites Conserveries. Paris: Dunod.

Killeit, U. 1987. The use of ascorbic acid and ascorbate in fruit and vegetable processing. Hoffman-La-Roche AG. Technical Bulletin.

Magee, T.R.A. and Wilkinson, C.P.D. 1992. Influence of process variables on the drying of potato slices. In Int. Journal of Food Science and Technology. 27, pp. 541-549.

MARRIOTT, J. and LANCASTER, P.A. 1984. "Bananas and plantains". In Handbook of Tropical Foods, pp. 85-144. H.T. Chan, Jr. (Editor). New Yorlc/Basel: Marcel Dekker.

MATZ, S.A. 1984. Potato Chips. In Snack Food Techology. AVI Publishing Co., Inc. Westport, Conn.

MEURER, P. and GIERSCHNZER, K. 1992. Occurence and effect of indigenous and external microbial enzymes in lactic acid fermented vegetables. In Acta Alimentaria, Vol.21 (2),pp.171 - 188.

PEDERSON,C.S. 1963. Processing by fermentation. In Food Processing Operations, Vol. 2. M.A. Joslyn and J.L. Heid (Editors). AVI Publishing Co., Westport, Conn.

POTTER, N.N. 1984. Food Science. 3rd Edition. AVI Publishing Co. Westport, Conn.

PRUTHI, J.S. 1963. Physiology, Chemistry and Technology of Passion Fruit. In Advances in Food Research, Vol. 12, pp. 203-274. Academic Press, Inc., New York, N.Y.

STAFFORD, A.E. 1984. "Mango". In Handbook of Tropical Foods, pp. 399-432. H.T. Chan, Jr. Editor. New York/Basel: Marcel Dekker

TAUFIK, Y.C.M. and KARIM, M.N.A. 1992. Storage stability of ciku leather. In ASEAN Food Journal, Vol. 7, No. 1, pp. 53-55.

THOMPSON, A.K. 1989. Recent advances in post-harvest technology of fresh fruits and vegetables. Private communication. Cranfield Institute of Technology, Silsoe College, U.K.

TORREGIANI, D. 1993. Osmotic dehydration in fruit and vegetable processing. In Food Research International. 26. pp. 59-68.

UTA. 1994. Deshydratation - La maîtrise technique. U.T.A. Industrie. Technical Bulletin. UTA: Cancon.

VAN ARSDEL, W.B. and COPLEY, M.J. (Editors). 1964. Food Dehydration. Vol. II Products and Technology. AVI Publishing Co., Westport, Conn.

VMC. 1994. Share our ideas as well as our glass. Technical Bulletin. VMC: Reims.

